An Introduction to Narrative Generators

An Introduction to Narrative Generators

How Computers Create Works of Fiction

Rafael Pérez y Pérez
and
Mike Sharples

OXFORD
UNIVERSITY PRESS

OXFORD
UNIVERSITY PRESS

Great Clarendon Street, Oxford, OX2 6DP,
United Kingdom

Oxford University Press is a department of the University of Oxford.
It furthers the University's objective of excellence in research, scholarship,
and education by publishing worldwide. Oxford is a registered trade mark of
Oxford University Press in the UK and in certain other countries

Published in the United States of America by Oxford University Press
198 Madison Avenue, New York, NY 10016, United States of America

British Library Cataloguing in Publication Data
Data available

Library of Congress Control Number: 2023932533

ISBN 978–0–19–887660–1
ISBN 978–0–19–887661–8 (pbk.)

DOI: 10.1093/oso/9780198876601.001.0001

Printed and bound by
CPI Group (UK) Ltd, Croydon, CR0 4YY

Links to third party websites are provided by Oxford in good faith and
for information only. Oxford disclaims any responsibility for the materials
contained in any third party website referenced in this work.

To Pilar, Rafael, and Yani, who are indispensable and always present.
To Susana, my life partner.

To Minji, Jenny, and Evelyn for their love and inspiration.

Preface

We first discussed writing a book on automatic narrative generation in 2001. At that time, the tentative title was *'Creativity in Computers: Computer-Based Storytellers'*. We estimated that it would be 150 pages long and would be finished by 31 May 2002. Different circumstances forced us to postpone the project.

Three years after that first attempt, we published an article which analysed and compared in detail the main characteristics of three story-generating systems (Pérez y Pérez and Sharples 2004). This work, in a way, functioned as a preview or prototype, exploring how we could materialize our interest in producing a book.

In February 2017, at a meeting in México City, we spoke again about the importance of resuming this project and agreed that, as soon as we finished our current commitments, we would start working on the manuscript. Finally, in September 2019, the necessary conditions were in place, and sitting at the kitchen table of a beautiful London apartment, we developed a roadmap to realize our long-standing ambition of writing a text on computer systems that generate narratives, which culminated in this book.

Many events have shaped the structure, writing style, and content of this work. Here, we share some of them with you.

During the late 1970s and until the beginning of the current century, the predominant vision in the automatic generation of narratives was the methodology known as planning. For those interested in learning about the subject, following details of the planning processes that occurred within these systems was complicated. Perhaps for this reason Meehan, one of the pioneers of story generation, begins a chapter, where he explains in detail how his system works, as follows:

> For the benefit of those who wish to see the innards of TALE-SPIN, I present this chapter. I'm going to follow the generation of a story from start to finish, giving as much detail as I can. I'm also going to describe the related sources of information, following, in effect, all the alternate paths in the maze as well as the correct one. Be warned that you may lose your bearings. Only computers are immune to such headaches.
>
> (Meehan 1976, p. 157)

Meehan was correct! The amount of information these systems handle when developing a story makes analysis of their inside working complicated. This situation led us to ask ourselves, is it possible to explain in detail but in an accessible way how these models work, especially for those new to the field? Our answer was yes, it is possible, but it will not be easy!

In 2011, during the International Conference on Computational Creativity held in México City, a group of three doctoral students from the United States explained to us that they were working on the automatic generation of narratives, but were concerned about how they could make a scientific contribution to the field. This situation caught our attention because, although interesting advances have been made, the limitations of narrative generation systems are evident. The conversation revealed that, for some students and researchers, it is not enough to have access to literature on the subject. Creating a coherent discourse on the scope and limitations of the main systems that have been developed over the years requires experience. For this reason, it seemed pertinent to write a book that allows those interested to explore some of the systems that have significantly influenced the field, and thus have the possibility of comparing them from different perspectives. We are convinced that this will contribute to the development of this area.

The advent of language models based on deep neural networks, especially from the development of the so-called transformers, has caused much discussion, both in the scientific community and in society in general, about the effects that AI will have on humanity. The release in 2022 of ChatGPT and in 2023 of GPT-4, developed by OpenAI, is a recent example. Some of the most controversial issues include ethical aspects of these developments, their ecological consequences, the problems of bias demonstrated by many of these programs, the loss of jobs caused by these technological advances, the way in which easy access to these technologies will affect education, the economy, culture, art, etc. The variety of actors involved in these debates is enormous, both in their age range and in their background, as well as their vision of the risks and benefits of these developments, among other aspects. Undoubtedly, the exchange of opinions has been rich and beneficial. Unfortunately, negative effects have emerged as well, such as growing misinformation about what AI can and cannot do. Sometimes motivated by economic, political, or other interest, or simply ignorance, there are people who make inaccurate assertions about the capabilities of these systems. It is common to find statements and conjectures in major media outlets that do not correspond to reality (see some examples in Chapter 1). This causes confusion and misinformation. A well-informed society about how these models work has the possibility of influencing those decisions that affect them. Unfortunately, it is difficult to find texts written for non-specialists that clearly explain the mechanisms that control intelligent systems. This text aims to address this problem.

These, and other experiences from more than 20 years of research and development, have shaped the text that you are about to start reading. The main objective of the book is to inform students and researchers interested in the subject, regardless of their programming skills, about how systems for the automatic generation of narratives work. It will help if you have a good capacity for abstraction, as well as skills in analysing logical processes. Throughout the book we explain what we consider to be some of the most representative techniques in this area. However, it was impossible to include all of them. Likewise, we describe the main features of recognized computer programs to illustrate the concepts studied, although, much to our regret, some works that left their mark on the development of the area have not been included. To reach a larger audience, we have avoided using technical language as much as possible. We have also been especially careful not to assume any previous experience with the methods studied. Thus, some chapters' function is to introduce, through carefully designed examples, essential concepts to understand topics discussed in later chapters.

Little by little, automatic narrative generators are becoming more of a part of our daily lives. Every day it becomes more relevant to understand how these systems work, to understand their scope and limitations. That is why we are so happy to share this work with you. Let's start.

Rafael Pérez y Pérez
Mike Sharples

México City
Mike Sharples, London
January 2023

Acknowledgements

The first time we discussed writing this book was in 2001. Different circumstances led us to postpone this project. Throughout this journey, we have been lucky enough to meet people who have been central in completing this work.

We want to acknowledge all at the School of Cognitive and Computing Sciences (COGS) at the University of Sussex, for providing such a stimulating interdisciplinary environment where creativity and innovation flourished, particularly Margaret Boden, who inspired both creativity and rigour.

We are thankful to colleagues from the computational creativity community, like Nick Montfort, Pablo Gervás, Mark Riedl, Carlos León, and Tony Veale, and from the Universidad Autónoma Metropolitana, like Eduardo Peñalosa and Vicente Castellanos, whose brilliant work and different perspectives have been an inspiration for us.

We would like to thank the people who reviewed drafts of the manuscript and offered valuable advice, including Tony Veale, Pablo Gervás, and the anonymous reviewers chosen by Oxford University Press.

We are deeply grateful to our Publisher at Oxford University Press, Dan Taber, to Giulia Lipparini, Rajeswari Azayecoche and all the editorial team, for their support in taking the book to publication.

Especially, we thank our families—Susana, Yanitza, Rafael, Tomy, Tomás, Isi, Vicente, Miguel, Minji, Evelyn, and Jenny—for their constant support and encouragement during the writing of this book.

Contents

List of Figures

List of Tables

1
The power to narrate

1.1 The relevance of narrative for humans

What are the most significant episodes in your life? Perhaps, how you fell in love for the first time, or what led you to meet the love of your life; your child's first great accomplishment; that unforgettable trip to unknown and spectacular territories; or how you got that brilliant idea that led to your main professional achievement. Or possibly, the circumstances that surrounded the farewell of a loved one, or the resounding personal disaster from which you thought you would never recover. What all these experiences have in common is that we think about them and we share them with others as narratives. We make sense of the world, relive the past and picture the future, in terms of narratives. We live through stories.

There are multiple studies that illustrate the fundamental function of narratives for the development of our cognitive and social skills. For instance, Ferretti et al. (2017) propose that human language evolved from storytelling. Early humans produced narratives by acting out sequences of past events, probably employing grunts, roars, and cheers to emphasize relevant information. Over time, language has evolved to express those acted accounts. Ferretti and colleagues show how mental time travel (MTT), that is, the human capacity to communicate past events and foresee the future, is essential to narrative production. MTT connects narrative with other cognitive processes, such as the role of emotions in making choices.

Some researchers argue that decisions about how to respond to a given situation rely on the emotions that we expect a particular response will elicit. They claim that the human ability of picturing the emotions that decisions would evoke, known as affective forecast, occurs as a consequence of the human capacity to relive personal events (Quoidbach & Dunn 2013). Similarly, Gelernter (1994) describes emotions as the glue of ideas during the creative process. These suggest that emotions play a fundamental role in the decisions that authors make during the development of a narrative. Other works have analysed the role of narratives in education. Mar et al. (2021) reported a survey of thirty-seven studies that compared learning either from stories or from explanations; their study suggests that learning through storytelling has a clear advantage from expository texts. Further studies focus on how narratives shape the behaviour of social groups. For example, Smith et al. (2017) examined the role of storytellers in nomadic communities in the Philippines. They found that the most-told tales were about how to collaborate and assist others in the community, for example, narratives about equality, tolerance, and friendship. In these societies,

An Introduction to Narrative Generators. Rafael Pérez y Pérez and Mike Sharples, Oxford University Press.
© Rafael Pérez y Pérez and Mike Sharples (2023). DOI: 10.1093/oso/9780198876601.003.0001

those camps with a greater proportion of storytellers were the most cooperative. In an analysis made by the researchers, storytellers were among the most valued members in the group, even ahead of skilled hunters and fishers.

In an interesting article published in *The New York Times*, Max Fisher[1] argues that nowadays there is a propensity among some politicians around the world to manipulate collective memory for their own ends. These leaders are replacing or reframing parts of the official history with new narratives that fulfill better their own agenda. In particular, this trend has seen the development of stories that exploit people's fears, real or imagined, and that articulate ideas closer to what their supporters want to hear to deal with those alleged threats. Fisher suggests that people are willing to listen to any leader that offers a version of history that, in their view, defends important beliefs that have been challenged and that appeal to their emotions. The article illustrates the author's points with examples in Russia, Hungary, China, USA, India, Poland, and the Netherlands. But it is easy to find similar situations in many other countries.

These are just few examples that illustrate the relevance of narrative for humans. Unfortunately, we know little about the mechanisms that make it possible to relate our experiences and thoughts. That is why we must make use of all the resources at our disposal to better understand this phenomenon. Computer science, and in particular artificial intelligence, is a powerful tool that can contribute to this endeavour. This book describes how computer programs can generate narratives and how studies of computational narrative can illuminate how humans tell stories.

1.2 Why develop computer models of narrative generation?

Because computer models provide a means for testing hypotheses and theories about the generation of narratives, we can think of them as tools for reflection on ideas. Let us elaborate this point. Sometimes, computer-based story generators characterize a relevant feature of a given theory. For instance, as we mentioned earlier, Ferretti et al. (2017) suggest that verbal language is not essential for storytelling. In this book, we will study some systems that instantiate this idea. This type of program creates a pre-linguistic representation of narrative based on data structures for concepts such as story plot and the goals of an author and characters. When the program generates the plot, it uses a variety of mechanisms to transform the data structures into text. Thus, the analysis of such programs might be useful for reflecting on some of the features, scopes, and limitations of the theory suggested by Ferretti and colleagues. These programs can be test-beds for investigating Ferretti and colleagues' idea that complete narratives can be constructed before they are turned into verbal language.

Similarly, we will study story generators that characterize concepts like those expressed by Gelernter (1994) and Quoidbach & Dunn (2013), that is, automatic

---

[1] https://www.nytimes.com/2022/01/05/world/history-revisionism-nationalism.html. Consulted on 21/Jan/2022

narrators that employ (computational representations of) emotions to progress a plot. We will also study narrative generators based on language models, theories of narratology, writing, daydreaming, and creativity, among others. In all these cases, computer-based narrators are useful tools for reflecting on ideas because:

- Theories used as frameworks to build computer models of narrative generation need to be expressed in great detail, forcing scientists to think about even the smallest aspects of those proposals. Thus, it is necessary to provide exhaustive descriptions of the features and functions of all components of the model. Because other approaches do not require this level of detail, this methodology provides a singular perspective of the phenomenon under study.
- Computer models of narrative generation are tested as running computer programs. As a result, it is possible to observe how the different components that form the system interact, and how the system behaves when one or more of those components are modified or disabled. This provides multiple perspectives on the process we are representing. This is one of its most powerful features.
- Computer programs generate outputs, in our case narratives. The features of the generated narratives provide elements for evaluating the model. For instance, if a group of human judges evaluates a given output as a well-composed story, it is possible to claim that, at least partially, the program fulfils its purpose (different programs might pursue different goals). Similarly, it is possible to observe how modification of some of the components of the system changes the characteristics of the generated narratives. We can run hundreds or even thousands of tests to try different configurations of the program in few seconds.

So, whether you want to represent aspects of a well-known theory or try out your own ideas, computer models for narrative generation offer an unlimited number of possibilities. Researchers figured out a long time ago the enormous potential of these tools. Thanks to the great work by James Ryan (2017) into the history of narrative generation, we know that the first computer program that generated short stories was developed by the linguist Joseph E. Grimes around 1960–1, when he was collaborating at Universidad Nacional Autónoma de México. He employed computers to study Mexican Indian languages like Hiuchol, spoken among indigenous peoples in the Sierra Madre Occidental (see Figure 1.1). Employing Vladimir Propp's narrative elements (Propp 2010), Grimes' system generated short stories in English and Spanish. The author claimed that the program could produce hundreds of thousands of variations on a single fairy tale theme. Grimes envisioned his story generator as a research instrument:

'These stories are an experimental tool,' says Dr. Grimes. 'I present the simulated stories to a native user of a language such as Senor Diaz and observe his reactions to them, looking especially for places where he bogs down in trying to follow the plot. Such observations on a number of stories, and a number of different native speakers, lead to a picture of the linguistic process at work. From these observations it is possible to proceed to a hypothesis about the underlying linguistic system.'[2]

Figure 1.1 Linguist Joseph E. Grimes working with Huichol Indian Silvino Díaz

Reprint Courtesy of IBM Corporation © 1963 (IBM, the IBM logo, and ibm.com are trademarks or registered trademarks of International Business Machines Corporation, registered in many jurisdictions worldwide.)

[2] 'Exploring the fascinating world of language', *Business Machines* 46(3), 10–11 (1963), courtesy IBM Corporate Archives.

Grimes was also aware of the capacity of computer models to assess theories (Grimes 1965). In a personal communication with Ryan, he mentioned of his work, 'In a way it was a test of Propp's hypothesis' that stories conform to an underlying narrative structure (cited in Ryan 2017).

In 1960, a MIT system called SAGA II developed scenes for television screen-plays in the cowboy/western genre that were showcased by CBS; and in 1968, Robert I. Binnick developed his own Proppian story generator (Ryan 2017). Some years later, based on Propp's work, Lakoff (1972) developed what became known as story grammars. Story grammars see stories as linguistic objects, which have a constituent structure that can be represented by a grammar. Although story grammars were created as a theory of story understanding, they can also be employed to generate new stories.

A grammar consists of a set of rewrite rules. Symbols that can be rewritten are called non-terminals; symbols that cannot be rewritten are called terminals. The following is an example of part of a story grammar developed by Thorndyke (1977, cited in Black and Wilensky 1979):

Story → Setting + Theme + Plot + Resolution
Setting → Characters + Location + Time
Theme → (Event)* + Goal
Plot → Episode*
Episode → Subgoal + Attempt + Outcome
Attempt → Event* or Episode

Each line of a story grammar is a single rewrite rule. It works as follows:

- The first rule in the grammar above specifies that a Setting, a Theme, a Plot, and a Resolution form a story. The same set of rules can be employed to generate new stories: to generate a Story, first generate a Setting, then a Theme, then a Plot, then a Resolution.
- The second rule establishes that the Setting can be rewritten as three components: Characters, Location, and Time.
- Next, the Theme is formed by one or more optional Events followed by a Goal State. Observe the notation used in this rule: the parenthesis indicates that the Event is optional, and the asterisk symbol indicates that the Event can occur one or several times.
- This process continues until only terminal symbols are reached and the grammar has generated the outline for a story.

Story grammars, in the form of rewrite rules of the kind shown above, are not powerful enough to recognize all and only stories generated by human authors (humans are good at breaking rules to create effects). But a story grammar can form the basis of a story generation system, to produce outline plots that the program then modifies and expands into a complete narrative. Examples of story generation systems based on story grammars include GESTER (Pemberton 1989) and TRACERY (Compton et al. 2015). These are the origins of computer-based narrative generation (although the attempts to automatize the production of texts have a long history; see Sharples and Pérez y Pérez 2022 for an analysis).

This type of system can be useful in areas like education, art, media, and entertainment. It can be a tool for writers, and for teaching writing, to help authors continue when blocked, help children develop their writing abilities, including developing a language to talk about creative composition, and to explore their writing processes. There is a high probability that some of the news that you read this morning was automatically produced by computers. Or that the stories recounted by your favourite character in the video game you played last night were the product of an embedded narrative generator. The possibilities are endless. However, realizing narrative generation as a computational algorithm is a complex task that usually requires an interdisciplinary approach and a well-experienced team. This endeavour might present enormous challenges.

To illustrate this point, think about your daughter or little niece. When a girl plays at being a mother, a scientist, or a fire fighter, she creates a scenario in which she includes only those characteristics familiar to her and that she considers relevant to her goal. In other words, she constructs a partial representation of the world that allows her to explore that space as she likes. Incorporating too many properties can make the system too complex and unmanageable; on the other hand, considering a few of them can make it trivial. Therefore, it is necessary to find the right balance. In the described scenario, the girl is creating an abstraction of what happens in her environment, which allows her to experiment and have fun.

The design of computer models requires an analogous process that makes achievable the realization of the model. Imagining in computer terms a love scene, where situations like intrigue, jealousy, passion, and so on are fundamental, establishes a great challenge for researchers. Interdisciplinary teamwork becomes essential to achieve this type of goal. Only through the fusion of methodologies, ways of understanding the world, and evaluation mechanisms can we advance in the understanding of this phenomena (for an analysis of the characteristics of interdisciplinary work see Pérez y Pérez 2015a). Undoubtedly, the knowledge that emerges from the humanities, social sciences, and arts is essential for developing this type of research instrument.

1.3 Why write a book about automatic narrative generation?

We are living in an era where many of our everyday activities are related in some way or another to artificial intelligence (AI). So it is necessary to have well-informed citizens able to engage in critical discussion about the benefits and risks that AI offers to society. However, most available books in AI are either too specialized, usually designed for readers with solid backgrounds in computer science, or excessively general, where systems and techniques are superficially described. It is hard to find material designed for readers with no background in computer science but who, nevertheless, are interested in understanding the core processes underlying these systems. We refer to this phenomenon as the *AI knowledge gap*. This book attempts to contribute to filling the AI knowledge gap in the field of automatic narrative generation.

We also hope that this initiative will help to reduce the dissemination of misinformation about automatic storytelling. Here two examples. Over many years, researchers have noted how people tend to attribute social and cognitive abilities to intelligent agents. ELIZA, a conversational agent developed in the 1960s, is a well-known example. Joseph Weizenbaum, its creator, wrote: 'Some subjects have been very hard to convince that ELIZA (with its present script) is *not* human . . . Eliza shows, if nothing else, how easy it is to create and maintain the illusion of understanding, hence perhaps of judgment deserving of credibility. A certain danger lurks there' (Weizenbaum 1966). Some narrative generators seem to produce similar effects on people. We expect that a better understanding of how these programs work will aid to reduce this phenomenon.

Now that there is a generalized interest in AI, we find that mass media discussions lack in-depth knowledge of the topic. As a result, we might hear or read inaccurate and confusing statements. A classic example is the claim that AI can mimic the human brain. A computer program cannot mimic an organ whose functioning we are just beginning to understand. Neither can a computer program replicate a mind. Computer models can imitate some basic neural and cognitive functions; however, because both brains and minds are complex and poorly understood, we are far from being able to replicate them. Another typical confusion is the claim that AI can learn like humans. The learning process of an intelligent system is different to that performed by a human. So AI systems do not learn like humans do.

These imprecisions promote misinformation that, in some cases, leads to unrealistic scenarios associated with AI. We hope that filling the knowledge gap will help to reduce these misunderstandings.

1.3.1 About this book

This book describes computer programs that generate narratives, in most cases, in the form of short stories. Let us introduce the definitions of narrative and computer-generated story that we employ in this work. Following Prince (1980), a narrative is the representation in a time sequence of imaginary or factual events. A computer-generated story is a sequence of actions that introduces characters and setting and advances one or more key characters through activities within the setting to produce a plot. It must show a coherent and narrative flow between activities and settings over time; it must show an overall integrity and closure, for example with a problem posed in an early part of the text being resolved by the conclusion; and it should generate expectation by setting up difficulties faced by the characters or tension in the plot

(Pérez y Pérez and Sharples 2004). It should present this in a form that humans can understand, for example, as a story outline or a text.

The book introduces different methodologies for the development of computer models for narrative generation. Rather than trying to describe all possible ways a computer system might generate a narrative, the text focuses on some of the most relevant techniques employed over the past 60 years. Although we provide details of the different approaches studied, we avoid as much as possible the use of technical language. This decision is not to the detriment of students and researchers in computer science interested in automatic narrative generation who, we believe, will find the descriptions and examples provided in this book valuable. In most parts, we offer an introduction to relevant concepts related to automatic storytelling, followed by a description of a well-known computer program that illustrates how such concepts are employed. We describe the strengths and limitations of the programs we analyse. The word 'limitations' should not be given a negative connotation; it simply represents the frontiers that a model has achieved and, therefore, points out possible routes to follow in order to progress the field. All the programs described in this book show the narratives that they generate as texts. We recommend implementing the examples described throughout the book. If you do not know how to develop a computer program, you can use material such as cardboard boxes, crayons, paper, and so on, to represent the processes that illustrates the described methods; if you know a programming language, you could develop your own system. This practice will provide you with a deeper understanding of the models explained in this work.

Thus, this book makes it possible to compare some of the many ways that researchers have characterized the automatic generation of narratives, and to comprehend the core properties that distinguish this area of knowledge.

The book is organized as follows. Chapter 2 describes the use of templates. This method has been used in applications ranging from commercial programs, for example automatic news generation, to systems for creating pieces of electronic literature.

Chapters 3 to 5 explain the use of problem-solving techniques for narrative generation, in particular how characters' goals can drive the generation of a narrative. Chapters 6 and 7 expands these ideas by describing planning techniques and by illustrating how the inclusion of authors' goals makes it possible to produce more elaborate texts.

Chapter 8 illustrates the basic operation of programs that use statistical methods to generate texts. The concepts introduced here help to better understand the topics covered in the next two chapters. Chapters 9 and 10 describe the use of neural networks for the generation of narratives. Instead of explaining the mathematical models on which these systems operate, we provide a general description of the processes that they perform. Through an example, in Chapter 9 we illustrate how a single artificial neuron, the basic element of a network, works; we also introduce some essential concepts to understand how more complex systems operate. On this basis, in Chapter 10 we describe the general characteristics of a deep neural network.

In Chapter 11 we employ our own research to illustrate how cognitive models can be used as frameworks for the development of computer models for narrative generation. First, we introduce the engagement–reflection cognitive account of creative writing; then, we describe how we use these ideas to build our story generator called MEXICA, and provide details of its implementation.

Chapters 12 and 13 offer a description of seven computer programs for narrative generation. The purpose is to complement the perspectives offered by analysing the systems studied in previous chapters. In this way, we compare how different narrative designers have expressed their visions about the purpose and goals of using computers models to study and generate narratives.

Finally, Chapter 14 synthesizes the core features of the systems studied through the book and reflects on their impact on society.

We hope you will enjoy this book as much as we have enjoyed writing it. Let us begin!

2

Narrative templates

2.1 Introduction to narrative templates

We can picture templates as forms, with slots left blank to be filled in. For example, it is common to find material for teaching English that employs templates to show texts that help students to consolidate concepts, for example the correct use of verbs: 'This morning John _____ to the park.' There are endless possibilities. In this section we analyse those created from a known text.

An interesting, and sometimes amusing, use of templates is to produce variations of a given written piece. To take a well-known text and use it to create a template is also a useful exercise for exploring different aspects of language, such as to illustrate the importance of the words chosen by an author to describe a passage, to analyse how small modifications might change the whole intention of a paragraph, and so on. We illustrate this with a fragment of the novel *The Brothers Karamazov* by Fyodor Dostoyevsky:

> The elder's absence from his cell had lasted for about twenty-five minutes. It was more than half-past twelve, but Dmitri, on whose account they had all met there, had still not appeared. But he seemed almost to be forgotten, and when the elder entered the cell again, he found his guests engaged in eager conversation.
>
> (Dostoyevsky 2009)

The first step is to select which elements in this fragment will be substituted. We choose the characters in the passage, in this case the elder, Dimitri, and the guests, because then we can substitute diverse actors and observe the effects in the narrated scene. We also select the location described in the text, that is, the cell, to experiment with diverse environments. Finally, we pick the action that the guests perform because it describes an interaction between characters. The result looks as follows:

> The _____'s absence from _____ had lasted for about twenty-five minutes. It was more than half-past twelve, but _____, on whose account they had all met there, had still not appeared. But _____ seemed almost to be forgotten, and when _____ entered _____ again, _____ found _____ engaged in _____.

An Introduction to Narrative Generators. Rafael Pérez y Pérez and Mike Sharples, Oxford University Press.
© Rafael Pérez y Pérez and Mike Sharples (2023). DOI: 10.1093/oso/9780198876601.003.0002

The next step is to decide how to fill the blanks. We refer to each blank as a *field*, a common term in computer science vocabulary. In this case, we have nine fields: Field-1 to Field-9. Field-1, Field-5, and Field-7 represent the main character; Field-2 and Field-6 represent the location where the action takes place; Field-3 and Field-4 represent the secondary actor; Field-8 represents a group of characters and Field-9 represents the action that they perform. To produce a smoother text, Field-4 and Field-7 are substituted by a pronoun.

The information that we insert into each field might be arbitrary. However, to keep the coherence of the text, we need to apply two rules: (1) all substitutions must respect the type of the original content of the field, that is, characters must be substituted with other characters, locations must be substituted with other locations, and the like; (2) the content of all fields must be related. This goal is achieved by choosing themes and filling the blanks with words that make sense within this subject. If we think of a religious scene, the main character might be a cleric; so, Field-1 and Field-5 are filled up with 'the priest' and Field-7 with 'he'; the secondary character might be a nun, so Field-3 is substituted by 'Sister Mary' and Field-4 by 'she'; Field-2 and Field-6, the location, are replaced by 'the church'; Field-8 is changed by 'the flock' and Field-9 by 'fervent praying'. The final text looks as follows (underlined words show fields that have been filled):

The <u>priest</u>'s absence from <u>the church</u> had lasted for about twenty-five minutes. It was more than half-past twelve, but <u>sister Mary,</u> on whose account they had all met there, had still not appeared. But <u>she</u> seemed almost to be forgotten, and when <u>the priest</u> entered <u>the church</u> again, <u>he</u> found <u>the flock</u> engaged in <u>fervent praying</u>.

We can try the same exercise with a different theme, for example mystery tales, where detectives and suspects are usually present:

The detective's absence from the crime scene had lasted for about twenty-five minutes. It was more than half-past twelve, but Lady Sanders, on whose account they had all met there, had still not appeared. But she seemed almost to be forgotten, and when the detective entered the crime scene again, he found all the suspects engaged in a bitter exchange of accusations.

With some practice, it is possible to mix themes using the same template, or even combine different templates, to produce unique texts.

In summary:

- Templates represent texts. They comprise two main elements: fixed information and a set of blanks.
- The fixed information provides a rigid structure that works as a framework to help in the production of a coherent text.
- The blanks are known as fields; they are filled with data that change.

- The content of the fixed information, as well as the position and function of the fields inside the text, are decided by the designer of the template.
- A well-designed template can be used several times to produce messages with diverse content; however, they always will have the same fixed information. As a result, they are limited, although sometimes very effective, tools for communication.

2.2 Filling templates automatically

The next step is to design digital templates. Computers are good tools for searching on the Web and for storing and processing huge amounts of information. We can take advantage of these features to access a vast source of data and, in this way, produce a greater variety of texts. Thus, a project to automate the generation of narratives requires designing digital templates, gaining access to or creating a source of data, and employing a computer program able to manipulate those data in order to fill in the templates. In this section we present two different ways of exploiting the power of computers in the generation of narratives.

2.2.1 Templates without fixed parts

Templates without fixed parts are a variant of the structure that we have discussed so far. They consist only of fields, so the final text depends entirely on the data used to fill the template. To illustrate this, we have developed a digital piece called 'Our Favourite Cities' that describes in a nondeterministic way some emotions that our favourite cities in the world trigger in us. It is inspired by the piece 'Scenes from a Marriage' by Milton Läufer (2018). Table 2.1 shows a spreadsheet with five fields and the data we employ for this project. Each field can be instantiated with six possible values. The template for Our Favourite Cities indicates the order in which each field will be printed (each field is represented by its name written in square brackets):

Table 2.1 Data employed for the piece 'Our Favourite Cities'

	A	B	C	D	E
1	**Time**	**City**	**Verb**	**Action**	**Connector**
2	at times,	Barcelona	makes me	laugh.	Thus,
3	always,	México City	allows me to	love.	Because,
4	now and then,	London	inspires me to	enjoy.	Therefore,
5	periodically,	New York	prompts me to	fantasize.	Then,
6	continuously,	Buenos Aires	enables me to	dream.	So,
7	endlessly,	Tokyo	encourages me to	hallucinate.	Given that,

[Time] [City] [Verb] [Action] [Connector]

The piece works as follows: the data to fill each field are chosen at random between the six possible options; when all fields in the template have been instantiated, the program starts over again. So, we have an endless text. The final product might look as follows:

always, New York enables me to dream. Thus, at times, México City makes me laugh. So, continuously, Barcelona inspires me to enjoy. Then, now and then, London makes me hallucinate. Therefore, endlessly, Barcelona encourages me to laugh. Given that, at times, Tokyo prompts me to enjoy . . .

As a second example, we use the same technique to generate a story. This piece, named 'She Won', utilizes four different types of data: those for the introduction (see Table 2.2a), those for the development of the plot (see Table 2.2b), those for the climax (see Table 2.2c), and those for the resolution (see Table 2.2d). For the sake of clarity, we distribute these data among four different spreadsheets, but they all could be located in one. The template for this piece looks as follows:

[Introduction-character-1] [Introduction-action] [Introduction-situation]
[Development-adverb] [Development-character-2] [Development-reaction]
[Climax-adjective] [Climax-action] [Climax-antagonist]
[Introduction-character-1] [Resolution-outcome] [Resolution-closure-1]
  [Resolution-closure-2]

The piece works as follows: the data to fill each field are chosen at random between the three possible options; when all fields in the template have been instantiated, the program stops. Here are two examples of possible narratives produced by this program:

Beth understood that this could not wait anymore.
Unexpectedly, her mother attempted to persuade her.
Shaking, she confronted them.
Beth laughed and left. Her nightmares were gone! This was a victory!

Table 2.2 Data employed for the piece 'She Won'

a) Introduction

	A	B	C
1	**Introduction-character-1**	**Introduction-action**	**Introduction-situation**
2	Beth	understood	that today was the day.
3	Jenny	knew	that this could not wait anymore.
4	Laure	agreed	that she was not ready yet.

b) Development

	A	B	C
1	**Development-adverb**	**Development-character-2**	**Development-reaction**
2	Precipitously,	her husband	supported her decision.
3	Unexpectedly,	her mother	attempted to persuade her.
4	Therefore,	her best friend	decided to hide from her.

c) Climax

	A	B	C
1	**Climax-adjective**	**Climax-action**	**Climax-antagonist**
2	Shaking,	she confronted	them.
3	Confident,	she talked to	him.
4	Pale,	she went to visit	her.

d) Resolution

	A	B	C
1	**Resolution-outcome**	**Resolution-closure-1**	**Resolution-closure-2**
2	returned home excited.	She had faced her demons!	She was a winner!
3	laughed and left.	She had tackled her fears!	She was a fighter!
4	felt happy.	Her nightmares were gone!	This was a victory!

Jenny agreed that she was not ready yet.
Precipitously, her husband decided to hide from her.
Pale, she confronted him.
Jenny returned home excited. She had faced her demons! She was a fighter!

These stories are intentionally ambiguous. But, as we will see in the following section, templates are able to generate more elaborated and coherent narratives.

2.2.2 Automatic production of sports news

The automatic production of sports news has grown in recent years due to its commercial value. Sports data companies hire people to go to football/soccer matches and report in real time what is happening. Nowadays, players wear devices during a match that make it possible to record information about their performance, so that detailed statistics can be obtained about each one of them. Even the balls used during the game have embedded microchips that allow data to be obtained. Most of this information ends up on websites specialized in sports that we can employ to fill templates.

Here is a brief description of how association football (soccer) works for those of you unfamiliar with this game. Two teams, the host and the visitor, play against each other for 90 minutes; the match is divided in two parts that last 45 minutes each, with a 15-minute break between them. Each team has eleven players. The team that scores more goals wins. In written reports or TV transmissions, the name of the host team is reported first, then the name of the visiting team; for example the sentence 'The Reds team will play tomorrow against The Blues team' implies that The Reds is the host team and The Blues is the visiting team. Something similar occurs with the score, where the goals of the host team are reported before the goals of the visiting team. Thus, the sentence 'the final score was 0-1' means that the host team did not score a goal while the visiting team scored one goal, winning the game. The match is played in the host team's stadium.

Although the information provided by diverse sports-news sites might differ, most of them include at least: the names of the teams involved, the date of the match, the name of the venue, the score at the end of the first half, and the score at the end of the match. These data can be stored in a spreadsheet with eight fields that can be employed to fill what we refer to as the soccer-general-template (see Table 2.3).

The soccer-general-template looks as follows:

On [Date] the [Host-team] played against the [Visitor-team] in the '[Venue]' Stadium. At the end of the first half the score was [Host-score-FH]-[Visitor-score-FH]. At the end of the game the result was [Host-score-end]-[Visitor-score-end].

The following shows an example of a generated text:

On November 1 the Reds played against the Blues in the 'Reddish' Stadium. At the end of the first half the score was 3-0. At the end of the game the result was 3-3.

The same template can be used for different games. Let us imagine that we build a spreadsheet with the results of the games played during the weekend in the local league (see Table 2.4). A programmer can develop software that generates a report for each row. Figure 2.1 shows the generated reports.

Table 2.3 Data employed to fill the soccer-general-template

	A	B	C	D	E	F	G	H
1	Date	Host-team	Visitor-team	Venue	Host-score-FH	Visitor-score-FH	Host-score-end	Visitor-score-end
2	November 1	Reds	Blues	Reddish	3	0	3	3

Table 2.4 Data that represents games of the weekend

	A	B	C	D	E	F	G	H
1	Date	Host-team	Visitor-team	Venue	Host-score-FH	Visitor-score-FH	Host-score-end	Visitor-score-end
2	November 1	Reds	Blues	Reddish	3	0	3	3
3	November 1	Jaguars	Lions	Jungle	0	0	0	1
4	November 2	Chihuahuas	Beagles	Dog Park	1	0	2	0
5	November 3	Sailors	Explorers	Ocean	4	2	4	3

On November 1 the Reds played against the Blues in the "Reddish" Stadium. At the end of the first half the score was 3-0. At the end of the game the result was 3-3.

On November 1 the Jaguars played against the Lions in the "Jungle" Stadium. At the end of the first half the score was 0-0. At the end of the game the result was 0-1.

On November 2 the Chihuahuas played against the Beagles in the "Dog Park" Stadium. At the end of the first half the score was 1-0. At the end of the game the result was 2-0.

On November 3 the Sailors played against the Explorers in the "Ocean" Stadium. At the end of the first half the score was 4-2. At the end of the game the result was 4-3.

Figure 2.1 Four different reports employing the soccer-general-template

In a few seconds, a computer might produce reports about soccer matches around the world. In fact, with very few changes, it is possible to use the same spreadsheet to generate texts using templates in different languages. The following shows a text generated using a template written in Spanish and the data in Table 2.3:

En noviembre 1 los Reds jugaron en contra de los Blues en el estadio 'Reddish'. Al terminal la primera mitad el marcador era 3-0. Al término del encuentro, el resultado fue 3-3.

2.3 Enhancing templates: alternative texts, calculating data, contextual templates

The soccer-general-template can be employed to publish information in blogs or sport-news sites. However, readers quickly will notice the repetitiveness of the text and probably will get bored. A practice to produce more appealing descriptions is to employ alternative expressions in the fixed part of the template. For instance, the description 'at the end of the first half' might be substituted randomly with any of the following options: 'After 45 minutes', 'When the first half ended', 'At half time'. In this book, alternative texts are indicated in the template by an inventory of possible expressions, separated by slashes and surrounded by curly brackets, for example {At the end of the first half/After 45 minutes/When the first half ended/At half time}. In a similar way, the soccer report can be improved by choosing alternatives for the expression 'At the end of the game the result', such as {At the end of the game the result/The final score/When the match ended the score}. The final template looks as follows:

On [Date] the [Host-team] played against the [Visitor-team] in the '[Venue]' Stadium. {At the end of the first half/After 45 minutes/When the first half ended/At half time} the score was [Host-score-FH]-[Visitor-score-FH]. {At the end of the game the result/The final score/When the match ended the score} was [Host-score-end]-[Visitor-score-end].

Fields can be treated similarly. Because computers are good at performing arithmetic operations, templates can take advantage of this feature. For instance, the data inside the Date field can be expressed in different ways, depending on when the match was played and when the report is published. Thus, Date might be substituted with, for example, 'Today', 'Yesterday', 'Two days ago', or 'One week ago'. To achieve this, one must calculate the number of days between the publication date and the date of the match. The next step is to choose the right text based on this number: if the totalled number of days is equal to zero then the field Date is substituted with 'Today'; if the number of days is equal to one then the field Date is substituted with 'Yesterday'; if the number of days is equal to two then the field Date is substituted with 'Two days ago'; otherwise, the field Date is substituted with 'In the past'. To make it work, we need to eliminate the word 'On' from the template. Thus, using the data in Table 2.4, these changes would produce the texts shown in Figure 2.2.

There are much more data that can be obtained from a football match to generate more elaborated content. For instance, before the match starts, the line-ups are announced; thus, we can get the nationality and statistics of all participants. During the match, each time a team scores a goal, the name of the player and the minute when she or he scored are recorded; the same occurs when a player is shown a yellow or a red card; and the like. That information makes it possible to produce customized reports. For instance, Table 2.5 shows details about the goals scored in our hypothetical Chihuahuas versus Beagles match: for the Chihuahuas, during the

Two days ago the Reds played against the Blues in the "Reddish" Stadium. At the end of the first half the score was 3-0. At the end of the game the result was 3-3.	Two days ago the Jaguars played against the Lions in the "Jungle" Stadium. After 45 minutes the score was 0-0. The final score was 0-1.
Yesterday the Chihuahuas played against the Beagles in the "Dog Park" Stadium. When the first half ended the score was 2-0. At the end of the game the result was 2-1.	Today the Sailors played against the Explorers in the "Ocean" Stadium. At half time the score was 4-2. When the match ended the score was 4-3.

Figure 2.2 Four different reports employing alternative texts and arithmetic operations

Table 2.5 Scores in the hypothetical match Chihuahuas versus Beagles

	A	B	C	D
1	**Chihuahuas**		**Beagles**	
2	Ventura	35$'$		
3	González	80$'$		

first half Ventura scored at 35$'$ and, in the second half, González scored at 80$'$; the Beagles were unable to score.

By contrast with previous examples, in this case the number of entries in the spreadsheet in Table 2.5, that is, the number of players that score a goal, changes from game to game. As a result, it is impossible to determine in advance how many fields are necessary for a template that narrates this important part of the game. To solve this problem, computer programs employ what is known as *lists*. A list stores a variable number of elements that can be manipulated in different ways; for example, one can get the first element of the list, or the second, or the third, or a subgroup of elements, and so on. Thus, for each team we need two lists, one that includes the names of the players and one that shows the minute they scored:

List-Host-Players-that-scored = (Ventura, González)
List-Host-Minutes = (35, 80)

In this example, the lists of the visitor team are empty because they did not score:
List-Visitors-Players-that-scored = ( )
List-Visitors-Minutes = ( )

Now we can build a template that gets the first name from List-Host-Players-that-scored and the first minute from List-Host-Minutes, and prints these data in the report; next, it gets the second name from List-Host-Players-that-scored and the second minute from List-Host-Minutes, and prints these data in the report. This template looks as follows:

For the [Host-Team](, [List-Host-Players-that-scored] scored at [List-Host-Minutes]).

All fields inside a pair of parentheses are lists. The parentheses represent a loop, so each element in the lists will be printed, one at the time. The final text looks as follows:

For the Chihuahuas, Ventura scored at 35$'$, González scored at 80$'$.

No matter how many goals the host team scored, this template will always work. Because for this example we know that only the host team scored, it is possible to develop a template that, employing lists, describes this specific situation. It employs data from the spreadsheets in Tables 2.4 and 2.5.

[Date] the [Host-team] played against the [Visitor-team] in the '[Venue]' Stadium. {It was a party for the hosts/It was a celebration day for the hosts}! {The visitors did not score/The visitors could never find the back of the net}! For the [Host-Team](, [List-Host-Players-that-scored] scored at [List-Host-Minutes]). In this way, {at the end of the game the result/the final score/when then the match ended the score} was [Host-score-end]-[Visitor-score-end].

The final text looks as follows:

Two days ago, the Chihuahuas played against the Beagles in the 'Dog Park' Stadium. It was a party for the hosts! The visitors could never find the back of the net! For the Chihuahuas, Ventura scored at 35′, González scored at 80′. In this way, at the end of the game the result was 2-0.

Because this template only works in a specific situation—when the host team scored at least once and the visitors did not score—we refer to it as *contextual*. Thus, contextual templates are useful for narrating particular circumstances of a game.

2.4 Building contextual templates

The first step to building contextual templates is to have a clear understanding of the dynamics of the incidents to be reported in order to cover the different situations that might arise during a game. In soccer, there are three basic cases: (1) when the host team wins the match, (2) when the host team ties with the visitor team, and (3) when the host team loses the match (see Figure 2.3). Each of these situations can be studied deeper.

For instance, case 1, when the host team wins, might be split up into the following scenarios:

Case 1.1: The host team scores and then it never loses its advantage.
Case 1.2: The host team scores first, next the visitor team equalizes, and finally the host team scores the decisive goal(s).
Case 1.3: The visitor team scores first, next the host team equalizes, and finally the host team scores the decisive goal(s).
And so on.

The deeper the analysis, the greater the number of contextual templates that can be created. So, for example, we can split Case 1.1, where the host team scores first and it never loses its advantage, into more detailed cases:

Case 1.1.1: The visitor team never scores.
Case 1.1.2: The visitor team scores at least once.

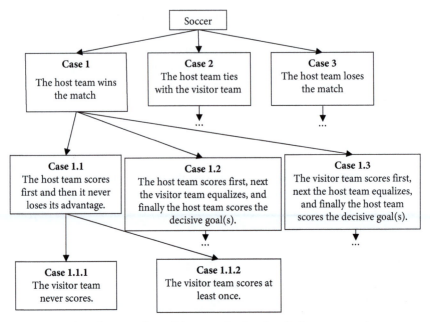

Figure 2.3 Classification of possible scenarios in a football soccer match

Figure 2.3 only shows some of the likely scenarios that might arise during a football match. With this information it is possible to build templates for each case and then mix them based on the particularities of each game:

Template general opening

[Date] the [Host-team] played against the [Visitor-team] in the '[Venue]' Stadium.

Template Case 1: The host team wins

The hosts {celebrated a victory/had a happy day}.

Template Case 1.1: The host team scores first and it never loses its advantage

{They were always ahead in the game/The team was always up in the score}.

Template Case 1.1.1: The visitor team never scores

{The visitors did not score/The visitors could never find the back of the net}!

Template case 1.1.2: The visitor team scores at least once

{Although they were able to find the back of the net, the visitors could never match the score/The visitors' goals were not enough}.

Template closing for case 1.1.1

For the [Host-Team](, [List-Host-Players-that-scored] scored at [List-Host-Minutes]). In this way, {at the end of the game the result/the final score/when then the match ended the result} was [Host-score-end]-[Visitor-score-end].

Template closing for case 1.1.2

For the [Host-Team](, [List-Host-Players-that-scored] scored at [List-Host-Minutes]). For the [Visitor-Team](, [List-Visitors-Players-that-scored] scored at [List-Visitor-Minutes]). In this way, {at the end of the game the result/the final score/when then the match ended the result} was [Host-score-end]-[Visitor-score-end].

Thus, in order to narrate the match between the Chihuahuas and the Beagles, we need to employ the following templates: general opening, case 1, case 1.1, case 1.1.1 and closing for case 1.1.1. The final text looks like this:

Two days ago, the Chihuahuas played against the Beagles in the 'Dog Park' Stadium. The hosts had a happy day. The team was always up in the score. The visitors could never find the back of the net. For the Chihuahuas, Ventura scored at 35′, González scored at 80′. In this way, at the end of the game the result was 2-0.

Employing the same set of templates, it is possible to relate the game between the Sailors and the Explorers (see Table 2.4). Table 2.6 shows the hypothetical scores for that game. Again, the host team beat the visitors; however, this time the visiting team scores. Thus, using the templates general opening, case 1, case 1.1, case 1.1.2 and closing for case 1.1.2, it is possible to generate the following:

Today the Sailors played against the Explorers in the 'Ocean' Stadium. The hosts celebrated a victory. They were always ahead in the game. The visitors' goals were not enough. For the Sailors, Smith scored at 15′, Mincioni scored at 20′, Ochoa scored at 30′, Peña scored at 44′. For the explorers, García scored at 25′, Cerezo scored at 35′, Piñeiro scored at 88′. In this way, when the match ended the result was 4-3.

Table 2.6 Scores in the hypothetical match Sailors versus Explorers

	A	B	C	D
1	**Sailors**		**Explorers**	
2	Smith	15′	García	25′
3	Mincioni	20′	Cerezo	35′
4	Ochoa	30′	Piñeiro	88′
5	Peña	44′		

The designers of a system like the one presented here determine in advance the structure of the final text, that is, the order in which templates are used. This ensures that the result is consistent. For instance, the soccer narratives shown in this chapter always include an introduction, a description of relevant events during the match, and a closure. Those are design decisions. Similarly, a method for analysing the available information is required, in order to choose the right templates for each game, that is, to determine which case in Figure 2.3 corresponds to the current match.

The variety and quality of the final text depends on how elaborated the templates are. The particularization of the templates relies on the available data and the type of analysis performed on them. Statistical analysis can be used to find trends, detect patterns, and so on. In this way, we can create templates for players that break records, show a low performance, tend to get tired soon in the game, have a tendency to increase their effectiveness in the second half, and so on. So templates and information analysis make it possible to generate sports stories that include heroes, villains, and victims. Designers sometimes rank templates based on their importance, in order to decide their position in the text and to choose between competing templates.

2.5 Final remarks about templates

The earliest recognized use of templates for story generations was a device known as the Movie Writer, patented by Arthur F. Blanchard in 1915.[1] The first known use of digital templates is almost as old as the general-purpose computer. Around 1952, Christopher Strachey employed templates to produced stylized love letters.[2] Strachey (1954) described his system in an article he wrote for the literary magazine *Encounter*.

One of the most complicated challenges in narrative generation is the automatic production of coherent and interesting events. Templates are useful in that sense because they provide a structure, created by humans, which guarantees those important requirements. That is why the use of this technique is popular with some companies dedicated to automatic text generation. Alternative texts, contextual templates, statistical analysis, and processes to calculate data can be employed to generate diversity in the generated narratives. The main limitation of this technique is its rigidity. A system based on templates will only generate narratives whose structure and content have been contemplated by the designer of the program. To add new templates might be a laborious task. The design of contextual templates that describes elaborated situations requires experienced designers.

In Chapters 3–5 we study one of the most popular techniques in narrative generation: problem-solving.

---

[1] John F. Ptak, 'Wood, Metal, Cubes, and Words: A Few Story-Machines since Gulliver's Travels', December 2015, Science Books, Post 2579. Available at https://longstreet.typepad.com/thesciencebookstore/2015/12/jf-ptak-science-books-quick-post-httpmentalflosscomarticle72112arthurblanchards-thinking-machine-was-random-story-ge.html, consulted 9 February 2022.
[2] Noah Wardrip-Fruin, 'Christopher Strachey: The First Digital Artist?', 1 August 2005. Available at https://grandtextauto.soe.ucsc.edu/2005/08/01/christopher-strachey-first-digital-artist, consulted 9 February 2022.

3
Narrative generation from the character's perspective

Problem solving

3.1 Introduction to problem-solving

For years, the study of how humans solve problems has taken the attention of researchers from diverse areas including psychology, pedagogy, cognitive science, and artificial intelligence. Much of the work in this area is based on the idea that human cognition always pursues an end; that is, it is aimed at reaching goals, and therefore, at finding ways to avoid or eliminate situations that prevent achieving those goals. These ideas have permeated the computer models of problem-solving that AI researchers have developed over the past 60 years. Most of those systems share three important features: a starting condition of the problem known as the initial state; the goal to be achieved known as the final or goal state; and a set of actions, known as operators, to convert a given state into a new one. Thus, solving a problem entails finding an appropriate sequence of actions that transforms the initial state into the final state. What makes an 'appropriate sequence' depends on the problem to be solved.

To illustrate these ideas, imagine you are visiting a city for the first time and decide to take the subway to go from your hotel to the National Museum downtown. You do not speak the local language, so you cannot get instructions, and the hotel has run out of maps. Because the city's subway system has several lines, there are different options to reach your destination. So the description of the initial state would be you standing in the train station in front of your hotel not knowing which direction to take; the description of the goal state would be you standing in the train station located near the museum. A possible sequence of actions to solve this problem is to choose one of the available directions that a train can take in the current station; get on the train to move to the next station; get out of the train either to change direction or to leave the station; and so on. To solve this problem, you must find the right sequence of actions that take you from your hotel to the museum. Since there are many train lines and you would need to change trains, then to find the solution by trial and error might take your entire holidays. That is why researchers in problem-solving have developed methods for reducing the complexity of this kind of situation. For example, if you know the general direction of the museum, it makes sense to take a train in that

An Introduction to Narrative Generators. Rafael Pérez y Pérez and Mike Sharples, Oxford University Press.
© Rafael Pérez y Pérez and Mike Sharples (2023). DOI: 10.1093/oso/9780198876601.003.0003

direction but only one or two stops then check again. Another method could be to look for signs, such as posters, at a station that might indicate the National Museum. In general, a good problem-solving strategy will combine general-purpose methods with ones specific to the problem.

These ideas are applicable to automatic narrative generation because tales include protagonists attempting to achieve goals and antagonists creating obstacles that get in the way of those protagonists. Some AI researchers have studied how to utilize problem-solving methods to build computer-based storytellers. In this way, a computer-generated story might be defined as the narration of the sequence of story-actions that characters perform to solve a story-problem to achieve a story-goal (the formal name used for these techniques is problem-space search). In this chapter, we study some important features of this type of automatic storyteller.

3.2 How problem-solving can be used to generate a narrative

Automatic storytellers use problem-solving techniques to build interesting narratives, for example heroes combating evil enemies who try to stop the main protagonist from achieving an important goal, villains chasing characters to hurt them, actors deciphering puzzles to discover a treasure. In all these cases, characters need to solve difficult problems. Usually, automatic storytellers break down troubles into subproblems that are easier to deal with. These, in turn, might be broken down into simpler difficulties. This approach is based on the idea that a collection of relatively simple challenges produces complex scenarios that result in interesting narratives.

We start this section with a short straightforward incident that is useful for introducing some relevant concepts. Picture a scene of a character lying in her bedroom in the middle of the night. Suddenly, she needs to escape from an intruder who breaks into her house. So she decides to lock herself inside the bathroom. Thus, to achieve her main goal of staying safe, the character first needs to solve the problem of going from her bedroom to the bathroom; then she must address the problem of locking the door to stay safe. The description of the sequence of steps necessary to accomplish these tasks make up the incident.

To build an automatic narrator, it is necessary to specify the features of the story-world. For this example, we can imagine a two-storey house. On the upper floor there is a bedroom which connects to a corridor that ends at the entrance of the bathroom. Thus, as Figure 3.1 shows, we can divide our story-world in four locations: *Bedroom*, *Beginning of the corridor*, *End of the corridor*, and *Bathroom*. This description provides a framework for designing the system.

Next, we need to figure out how to represent basic story elements like characters, story-actions, goals, and so on. Computers organize information around structures that the system can manipulate. For instance, to represent a character it is necessary

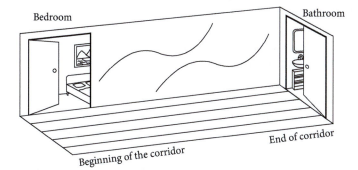

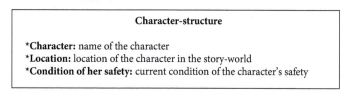

Figure 3.1 Story-world: Bedroom (1), Beginning of the corridor (2), End of the corridor (3), and Bathroom (4)

Character-structure

*Character: name of the character
*Location: location of the character in the story-world
*Condition of her safety: current condition of the character's safety

---

Figure 3.2 Representation of a character-structure

to define a structure, which we refer to as character-structure, and that is composed of a set of attributes. In this case, those attributes are **Character*, which is a string of letters or symbols to identify the actor in the incident; **Location*, which indicates the position of the character in the house, in this case *Bedroom, Beginning of the corridor, End of the corridor*, or *Bathroom*; and **Condition of her safety*, which can be set to either *Safe* or *In-danger* (see Figure 3.2), and that signals whether the final-state has been reached. The symbol '*' at the beginning of each attribute denotes that its value is assigned while the program is running rather than be predefined by the designers, as we will see in the example that follows starting in Section 3.2.1. The starred information can be modified at any moment; each character in a story has its own structure.

For the sake of clarity, all attributes in Figure 3.2, and in all other figures in this chapter, are represented in an easy-to-read format that is not necessarily the way they would be represented in a programming language.

3.2.1 Representing story-actions

The story-actions (not necessarily in the right order) that can occur in this exercise are *Lock the bathroom door, Walk into the bathroom, Walk to the beginning of the corridor*, and *Walk to the end of the corridor* (in front of the bathroom).

The number and type of deeds depend on the story-world and the features of the problems to be solved. To solve the problem and achieve the goal, the character needs to perform the correct actions, in the right order. As illustrated in Figure 3.3, each story-action has the following attributes:

- *Name of the action.* This is a unique identifier.
- **Character.* The name of the character performing the action.
- *Preconditions.* An associated set of requirements which need to be fulfilled before the action can be performed.
- *Post-conditions.* A set of consequences which describe the effects in the story-world after performing the deed.

In this example, we only have one character, but a story might have several actors. The *Name of the action* is an identifier that is set in advance and cannot be modified while the program is running. The value of **Character* is defined at runtime and can be changed at any moment; that is, different actors can perform the same action (in computer-science terminology, the expression 'runtime' is used to describe the moment when a program is running). *Preconditions* and *Post-conditions* cannot be modified while the program is running. For example, Figure 3.3c shows that to satisfy the precondition *Walk into the bathroom*, it is first necessary to be in front of the bathroom. The post-condition of *Walk into the bathroom* is that, as a result of performing the action, the character is now inside the bathroom. This is the common-sense knowledge that needs to be specified to the computer. Thus, for our program, performing an action entails modifying the value of one or more attributes.

(a) Story-action structure	(b) Story-action structure
Name of the action: Walk to the beginning of the corridor ***Character:** Name of the character performing the action **Preconditions:** *Character must be in the bedroom OR *Character must be at the end of the corridor **Post-conditions:** *Character is at the beginning of the corridor	**Name of the action:** Walk to the end of the corridor ***Character:** Name of the character performing the action **Preconditions:** *Character must be at the beginning of the corridor OR *Character must be inside the bathroom **Post-conditions:** *Character is at the end of the corridor
(c) Story-action structure	(d) Story-action structure
Name of the action: Walk into the bathroom ***Character:** Name of the character performing the action **Preconditions:** *Character must be at the end of the corridor **Post-conditions:** *Character is inside the bathroom	**Name of the action:** Lock the bathroom door ***Character:** Name of the character performing the action **Preconditions:** *Character must be inside the bathroom **Post-conditions:** The condition of the *Character's safety is set to 'Safe'

Figure 3.3 Four story-action structures

Figure 3.3a shows two preconditions for the action *Walk to the beginning of the corridor*: the character must be located in an adjacent location, that is, either inside the bedroom or at the end of the corridor. A similar situation happens when the character wants to walk to the end of the corridor (see Figure 3.3b): she must be situated either at the beginning of the corridor or inside the bathroom. These two preconditions are included to cover all possible situations to perform the deed (including ones that may not be used for this incident). Finally, notice that the attribute **Character* in the story-action's structure is linked to *Preconditions* and *Post-conditions*. That is, the character mentioned in the description of the preconditions and the post-conditions in Figure 3.3 refers to **Character*. Thus, to perform an action, the first step is to assign a value to this **Character* attribute.

3.2.2 Representing goals

After we have defined the set of story-actions, now we focus on the representation of goals. As mentioned earlier, the character first needs to solve the problem of moving to the bathroom; then she can sort out the problem of locking the door to stay safe. As shown in Figure 3.4, the attributes of the structure that represents the goal of moving to the bathroom are:

- *Name of the goal.* An identifier that cannot be modified while the program is running.
- **Character.* The name of the character involved in accomplishing the goal; the value of this attribute is assigned at runtime.
- *Preconditions.* The requirements that need to be fulfilled before the system can work on the plan. Preconditions are optional.
- *Plan.* The sequence of story-actions to be performed to achieve the current goal. The inclusion of a plan is compulsory.

This goal has no preconditions. Its plan presumes that, when the program starts, the characters is in her bedroom. This makes sense because the description of the

---

Goal-structure

Name of the goal: Moving the character into the bathroom
*Character: Name of the character involved in achieving the goal

Preconditions: none
Plan:
*Character performs the action 'Walk to the beginning of the corridor'
*Character performs the action 'Walk to the end of the corridor'
*Character performs the action 'Walk into the bathroom'

Figure 3.4 Representation of the goal *Moving the character into the bathroom*

incident indicates so. However, later we will modify the system so it can work with different initial locations. If one action in the plan cannot be performed because its preconditions are not fulfilled, the whole plan fails and, therefore, the goal is not achieved.

Next, we represent the goal of locking the door to stay safe. Figure 3.5 shows that the precondition of this goal is another goal. That is, in order to achieve the goal *Lock the door to stay safe*, first the character must accomplish the goal of being situated inside the bathroom. Then she can go ahead with the plan, in this case, to perform the action *Lock the bathroom door*.

Goal-structure

Name of the goal: Lock the door to stay safe
***Character:** Name of the character involved in achieving the goal
Preconditions: GOAL: Moving the character into the bathroom
Plan:
*Character performs the action 'Lock the bathroom door'

Figure 3.5 Representation of the goal *Lock the door to stay safe*

Figure 3.6 shows what we call a goal-oriented plot graph (GOP-graph). This graph displays the goals in the plot, their preconditions, and their plans (cf. Figures 3.4 and 3.5). Descriptions in bold represent goals; the rest represent either story-actions or situations that need to be fulfilled (e.g. that a character is in the bathroom). Arrows pointing towards a goal represent preconditions; arrows going outside a goal represent plans. The goal at the top of the map is the one that drives the development of the plot, known as the *driving goal*. Thus, the GOP-graph depicts the relations between the core components of the incident.

Now we can develop a computer program that narrates our incident. The system must be able to represent the structures and to execute the plans we have just designed; as an outcome it produces a text that narrates the steps that the character follows during her efforts to achieve her goal. It is beyond the scope of this book to explain how this program can be implemented in a specific computer language; instead, we describe how it works in general.

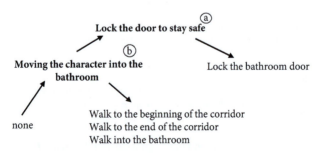

Figure 3.6 The goal-oriented plot graph of the incident

3.2.3 A computer program that generates an incident about staying safe

To start the narrative generation process we need some information: the name of the character for this incident, and her initial state. For this exercise, we have decided to employ default settings (see Figure 3.7). In this case, the name of the character is Jenny, she is in her bedroom, the condition of her safety is in danger, and the driving goal is to lock the bathroom's door to stay safe. These initial values match the description of the incident at the beginning of Section 3.2.

```
                        Default settings

    Name of the character: Jenny
    Initial location: Bedroom
    Initial condition of her safety: In-danger
    Driving goal: Lock the door to stay safe
```

Figure 3.7 Default settings to build the initial state

Employing this information, the system performs the following steps:

(1) The program creates the character-structure shown in Figure 3.8 that represents Jenny's state in the story-world; this is the initial state. Each time a story-action is performed, the state of Jenny changes. The program will end when the driving goal is reached, that is, when the character-structure's attribute *Condition of her safety* is set to *safe*. Thus, the different states in the story-world are represented by the values of the character's location and the condition of her safety.

(2) The program looks in memory for a goal-structure that matches the driving goal. If it cannot find one, the program fails and the incident cannot be written. In this case, it comes across the goal-structure shown in Figure 3.5.

(3) The system registers that Jenny is the character trying to lock the bathroom's door to be safe. That is, the system assigns Jenny to the attribute *Character* in the matched goal structure.

(4) The next step is to satisfy the active goal's preconditions (see Figure 3.6a). The precondition of locking the door to be safe is to achieve the goal of moving Jenny into the bathroom. So, now the program puts on hold the active goal of locking the door and instead focuses on how to move the character into the bathroom.

```
                    Character-structure

        *Character: Jenny
        *Location: Bedroom
        *Condition of her safety: In-danger
```

Figure 3.8 Representation of the initial state

3.2.4 Focusing on moving the character into the bathroom

(5) The system looks in memory for the goal-structure *Moving the character into the bathroom* shown in Figure 3.4 and assigns the value Jenny to the goal's attribute *Character to indicate that she is the one trying to move to the bathroom (see Figure 3.6b).

(6) Figure 3.4 indicates that the active goal has no preconditions.

(7) Next, the program focuses on executing the plan shown in Figure 3.4.

(8) The first action in the plan is to locate Jenny at the beginning of the corridor. The system gets the action-structure named *Walk to the beginning of the corridor* shown in Figure 3.3a, and sets *Character to Jenny to indicate that she is the one performing the action. Because Jenny is already in her bedroom, the precondition is fulfilled and, therefore, the post-condition can be applied. So the program moves Jenny to the corridor (i.e. it modifies the attribute *Location in Jenny's character-structure to *Beginning of the corridor*). Because she is in a new position, Figure 3.9a shows how the current-state in the story-world has been updated.

(9) Next, the system attempts to perform the second action in the plan for moving Jenny to the bathroom. The program looks in the group of actions shown in Figure 3.3 for the structure named *Walk to the end of the corridor*. The action's attribute *Character is set to Jenny and, because one of the preconditions is satisfied, the post-condition is triggered and now Jenny is at the end of the corridor (i.e. Jenny's character-structure is updated with the new position as shown in Figure 3.9b).

(10) The system performs the last action in the plan. The program looks into the group of actions shown in Figure 3.3 for the structure named *Walk into the bathroom*; then *Character is set to Jenny and, because the precondition is satisfied, the post-condition is triggered and now Jenny is inside the bathroom (i.e. Jenny's character-structure is updated with the new position as shown in Figure 3.9c).

So, the goal *Moving the character into the bathroom* has been achieved. Now the program returns to deal with the goal of locking the door to stay safe.

(a) **Character-structure**	(b) **Character-structure**	(c) **Character-structure**
Character: Jenny *Location:* Beginning of the corridor *Condition of her safety:* In-danger	*Character:* Jenny *Location:* End of the corridor *Condition of her safety:* In-danger	*Character:* Jenny *Location:* Bathroom *Condition of her safety:* In-danger

Figure 3.9 The current-state in the story-world after performing each of the three actions to achieve the goal of moving to the bathroom

3.2.5 Focusing again on achieving the goal of locking the door to stay safe

(11) Because Jenny is already in her bathroom, the precondition of locking the door of the bathroom to stay safe is satisfied (see Figure 3.5) and now the plan to achieve this goal can be executed (see Figure 3.6a).

(12) The program looks into the group of actions shown in Figure 3.3 for the structure named *Lock the bathroom door*; then the action's attribute *Character* is set to Jenny and, because the action's precondition is satisfied, the post-condition is triggered and Jenny locks the door (i.e. the attribute *Condition of her safety* in Jenny's character-structure is set to *Safe*).

(13) The program has achieved the character's goal and now it prints, using templates (see Chapter 2), the sequence of actions performed: Jenny wants to lock the bathroom door to stay safe. She goes to the corridor, walks through it and goes into the bathroom; then she locks the door. The end. We will describe below how to improve this version. This brief incident provides a context that allows reflecting on some important concepts related to problem-solving in storytelling.

3.3 Characteristics of a narrative generation system based on problem-solving

Problem-solving narrative generators, like the one presented here, work under the assumption that characters have goals that can be represented as problems to be solved. The sequence of actions performed to sort those difficulties out compose the narrative. Thus, goals are the backbone of any narrative generation system based on problem-solving. As a result, when building an automatic storyteller, important design decisions depend on them. The following discussion elaborates these ideas.

3.3.1 Coherence

The use of goals helps to produce in the reader the illusion that characters have motivations. In other words, because all performed actions aim to solve the main challenge in the tale, actors' behaviour seems to have a purpose and therefore the story is perceived as logical. All the goals that characters can accomplish, as well as their plans to achieve them, are established by the human designer of the system. In this way, because all the steps to deal with a problem have been defined in advance, the tale's overall coherence is guaranteed (although unexpected conducts of the actors may occur within the story).

3.3.2 Design of the system

The story-world, the group of characters, their number and type of attributes, the set of actions to be performed, their preconditions and post-conditions, and so on, are also defined by the designer of the automated storyteller. The features of all those elements are connected to each other. There is a clear relation between the challenges to be pursued and the topics that the stories are about. The number and type of characters depend on them. For instance, narratives about the Old West require cowboys, a sheriff, bad guys, and the like; accounts about the Middle Ages need kings, knights, princesses, and the like. For these kinds of topic, we are probably not interested in knowing whether the actors have locked the door, so that attribute is not useful. Rather, we might like to know whether those individuals are alive or dead, so other attributes need to be considered. In a similar way, the objectives of a sheriff are different to those pursued by a knight (and to those aimed by Jenny).

The characteristics of the story-actions are strongly linked to the goals to be achieved. In Section 3.2, there are no actions that describe how Jenny gets up from her bed or calls her best friend in the morning because they are not needed to solve the problem. In a similar way, the story-world consisted only of one bedroom, one corridor, and one bathroom. The size of the bedroom, the colour of its walls, the type of bed and cabinets used to furnish it, and so on, are never established; the corridor and bathroom share similar limited descriptions. Because that information is not necessary to progress the tale, we do not represent it in computer terms (although they could be used to enrich the narrative).

The representation of story's states, that is, the initial state, the final state, and any intermediate state between these two, are vital for an automatic storyteller. They represent the current state of affairs in the story-world and, therefore, show when a goal has been achieved. In Section 3.2, we employed the character-structure to represent the story's states because it provides the necessary information to identify when Jenny has achieved her goal; however, in other contexts, it might be more convenient to employ a structure exclusively built for this purpose.

3.3.3 Operation of the automatic storyteller: instantiation of characters

All the structures presented in this chapter have a common attribute: *Character. As mentioned earlier, its value is assigned at runtime. This is how it works. When the program starts, it selects the name of the actor in the tale; in this case Jenny. The system employs a character-structure to represent this actor and assigns Jenny to the attribute *Character (see Figure 3.10a). From now onwards, each time the system needs information about the character, it will check this structure.

Now, when the storyteller requires that a goal be achieved, it is necessary to specify who is attempting to solve the problem. That is why the goal-structure

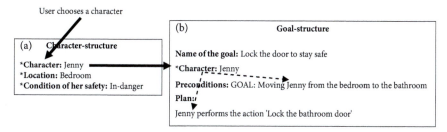

Figure 3.10 How the attribute *Character* is instantiated in other structures

also includes the attribute *Character. So the system copies the value Jenny from the character-structure's *Character into the goal-structure's *Character (see Figure 3.10). As a result, the program now knows that the precondition of the goal *Lock the door to stay safe* is a new goal: *moving Jenny into the bathroom*. In a similar way, the deed in the plan becomes *Jenny performs the action Lock the bathroom door.*

The system works similarly while processing other structures. Thus, if we modify our narrative to include more participants, the program can use the same goal-structures; it only must identify the actor pursuing the objective by assigning the correct value to the goal-structure's attribute *Character.

3.3.4 Backwards reasoning

In general, narrators based on problem-solving work backwards. Given a goal state, the system attempts to satisfy all its unfulfilled preconditions, each of which might have its own unsatisfied preconditions, and so on. This process continues until there are no more preconditions to satisfy. Then post-conditions start to be applied until the system comes back to the original goal in order to perform its plan. Preconditions and post-conditions might include diverse elements: one or more goals, specific values for some characters' attributes, and so on. There are multiple ways in which backward reasoning can be implemented; nevertheless, the core idea of putting on hold one process while the system focuses on solving other problems, and then coming back to continue working in the original process, is central for many storytellers based on problem solving techniques.

3.3.5 Diverse initial locations

One way to extend our program is to incorporate a routine that randomly chooses the initial location of our character, instead of always placing it in the bedroom. This way we would have more variety in the texts. However, if in the example in Section 3.2 the initial location is set at the beginning of the corridor or the end of the corridor (see Figure 3.7), the program will not be able to develop a narrative. This fault arises

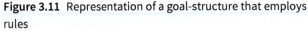

Goal-structure

Name of the goal: Moving the character from any position into the bathroom
*Character: Name of the character involved in achieving the goal
Preconditions: None
Plan:
If *Character is located in the bedroom then
 *Character performs the action "walk to the beginning of the corridor"
If *Character is located at the beginning of the corridor then
 *Character performs the action "walk to the end of the corridor"
If *Character is located at the end of the corridor then
 *Character performs the action "walk into the bathroom"

Figure 3.11 Representation of a goal-structure that employs rules

because the plan for moving the character into the bathroom (see Figure 3.4) presupposes that always the initial location will be the bedroom; as a result, the system is not prepared to deal with a different starting scene. One solution is to modify the goal-structure; so, as shown in Figure 3.11, we eliminate the precondition and include rules in the plan (cf. Figure 3.4). Rules have two sections and work as follows: when the conditions in the 'if' section are satisfied, the program executes the instructions defined in the 'then' section. Returning to our example, let us imagine that the system chooses *Beginning of the corridor* as the initial location. Employing the new plan, the system starts by verifying whether the character is in the bedroom; because the conditions are not satisfied, the action *Walk to the beginning of the corridor* is not performed. Next, the system checks whether the character is at the beginning of the corridor; this time, the condition is fulfilled and therefore the action *Walk to the end of the corridor* is performed. Finally, the system checks whether the character is at the end of the corridor; because the condition is fulfilled, the action *Walk into the bathroom* is performed. This solution works for any initial location that the system selects. Thus, the use of rules makes it possible to design more flexible plans. We have changed the name of the goal in Figure 3.11 to *Moving the character from any position into the bathroom* because this shows better the more general purpose of this structure.

3.3.6 Generation of the text

As a result of achieving the story's main goal, the system produces a sequence of story-action structures that describes how this problem is solved. Next, it is necessary to develop a mechanism that transforms those structures into readable text. One option is to include a new attribute into the story-action structure that describes the deed in English. For instance, for *Walk to the beginning of the corridor*, we can add the text 'Character reached the old corridor', where 'Character' is substituted by the value of the attribute *Character, in this case for Jenny; the action *Walk to the end* *of the corridor* is linked to the text 'Character walked through without paying much attention to the colourful walls'; for the action *Walk into the bathroom* the associated text is 'Character opened the brown door and came into the bathroom'; and for *Lock the bathroom door* the added text is 'Shaking, Character locked the bathroom door'. Now, the system can print the incident:

> *Jenny reached the old corridor. Jenny walked through without paying much atten-*
> *tion to the colourful walls. Jenny opened the brown door and came into the bath-*
> *room. Shaking, Jenny locked the bathroom door. The end.*

The system automatically inserts 'The end' after the last action in the tale. This text can be improved by developing some automatic mechanisms that replace some of the characters' names with pronouns. In this way, we can get the final version of this incident:

> *Jenny reached the old corridor. She walked through without paying much atten-*
> *tion to the colourful walls. She opened the brown door and came into the bathroom.*
> *Shaking, Jenny locked the bathroom door. The end.*

The order of that sequence can be modified; some techniques can automate this process. This is useful because different genres of narrative might require different descriptions: for example, a mystery tale usually starts with a murder case. So, employing the words 'because previously' as a connector, the story we generated might be presented to the reader in a different order:

> *Shaking, Jenny locked the bathroom door because previously Jenny reached the old*
> *corridor. She walked through without paying much attention to the colourful walls.*
> *She opened the brown door and came into the bathroom. The end.*

3.4 Final remarks about the introduction to problem-solving in storytelling

In this chapter we discussed how characters in narrative generators based on problem-solving have goals that can be represented as problems to be resolved, and how the sequence of actions performed to unravel those complications compose the narrative.

We studied how, to build an automatic narrator, it is necessary to specify the features of the story-world, and to represent story-actions and characters' goals. Then we described how these ideas can be used to produce a narrative. We also explained that narrators based on problem-solving work backwards. Finally, we showed how predefined texts can be used to present to the reader the narrative generated by the system. In Chapter 4 we describe more elaborated ways of producing narratives employing problem-solving techniques.

4

A methodology for designing a narrative generator based on problem-solving

4.1 Introduction to the methodology

This section introduces a methodology for developing a narrative generator based on problem-solving. Our approach starts with a very general idea, which is refined little by little, until we can specify the data structures to be used in the program. These data structures must be general enough to be employed in diverse narrative contexts. The methodology comprises four steps:

(1) Choose a main problem that a character must solve. This is known as the driving goal.
(2) Develop the story's outline where the driving goal is the core motivation.
(3) Using the outline as a reference, build a goal-oriented plot graph.
(4) Employ the goal-oriented plot graph as a guideline for defining the characteristics of the story-world and the attributes of the data structures.

This work is done by the human designer of the narrative generator. The resulting data structures are then coded into the story generator program which attempts to satisfy the goals, perform the plans, and report the actions as output text. In this section we describe steps one and two. The rest are elaborated in the following sections.

(1) Choose the driving goal. There are endless options that can work as a driving goal in a text. Abduction is a common theme in fiction. Thus, for this exercise, we will work with a narrative where the driving goal is *A hero decides to rescue a hostage*.
(2) Develop the story outline. Its purpose is to define the core events in a plot. In this case, our outline includes two main sections: those events that lead to a situation where the hero decides to rescue the hostage, and those events that lead to the liberation of the victim (see Figure 4.1). In the first section, we introduce the main characters in the tale and then provide a context that explains why later the hero decides to help the victim. This explicative context is useful for building a sense of coherence. Also, we include a situation where a kidnapper makes a character a hostage. In the second section, we incorporate a rescue scene that we have divided into two parts: (1) the hero learns,

An Introduction to Narrative Generators. Rafael Pérez y Pérez and Mike Sharples, Oxford University Press.
© Rafael Pérez y Pérez and Mike Sharples (2023). DOI: 10.1093/oso/9780198876601.003.0004

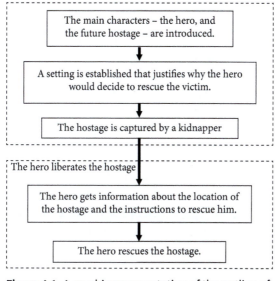

Figure 4.1 A graphic representation of the outline of the story

from a reliable source of information, the location of the hostage and how to rescue him; (2) the hero rescues the hostage. We have chosen this sequence of incidents based on our knowledge about abductions. You can find similar outlines in many well-known pieces of fiction. This is just one of many possibilities that the designer can select to build the system.

The next step is to employ the outline in Figure 4.1 as a reference to build a goal-oriented plot graph (GOP-graph).

4.2 Building a goal-oriented plot graph (GOP-graph)

As we explained in Chapter 3, the main purpose of a GOP-graph is to represent the plot as a graph organized around descriptions of the characters' goals (shown here in italic), their preconditions, and their plans for achieving the goals. We have already established that the plot includes two main parts: those situations that culminate in the hero deciding to take actions to liberate the hostage, and those incidents involved in the process of freeing the captive. The former works as the preconditions of the driving goal, and the latter works as its plan. To build the GOP-graph, we start by putting the driving goal on the top of the chart and then describing on the left its preconditions.

(1) DRIVING GOAL: *A hero decides to rescue a hostage* (description of its preconditions).

Based on Figure 4.1, the driving goal has three preconditions (see Figure 4.2a):

- Choose the characters that will play the roles of the hero and the hostage. Each time a tale is generated, the program might select different protagonists.
- Devise a reason to justify why the hero later decides to rescue the hostage.
- Indicate that the story requires an abduction scene. Because this is a relevant event in the plot, we represent it as a goal shown as *Kidnapping the hostage*. As a rule of thumb, important circumstances must be represented as goals to introduce new challenges to the protagonists at critical moments in the story.

Because we have included a new goal as part of the preconditions, next we focus on this new goal, and later we will return to define the plan for the driving goal.

(2) GOAL: *Kidnapping the hostage* (description of its precondition and plan).

This goal has one precondition for it to be satisfied, and its plan includes two actions (see Figure 4.2b):

- The precondition is that the character who plays the role of kidnapper should be specified. The system may choose a different kidnapper each time it generates a tale.
- The two actions are as follows: the kidnapper makes the hostage a prisoner, and then the kidnapper moves the hostage to a secret location.

The way we have organized this goal is based on our knowledge about abductions, and how we want to represent this scene. In this case, the kidnapping comprises only two actions. In later versions of this automatic storyteller, we might include a more elaborated sequences of events.

Now, we return to continue working on the driving goal. Because all its preconditions have already been defined, we concentrate on its plan.

(3) DRIVING GOAL: A hero decides to rescue a hostage (description of its plan).

- Based on Figure 4.1, and because this is a relevant event in the plot, we represent the plan of the driving goal as a new goal named *The hero liberates the hostage*.

Because we have included a new goal as part of the plan, now we focus on it.

(4) GOAL: *The hero liberates the hostage* (description of its precondition).

Based on Figure 4.1, the liberation scene has two parts: (1) finding out the location of the hostage and getting instructions to rescue him; (2) performing the instructions to free the captive. The former works as the goal's preconditions and the latter as its plan (see Figure 4.2c):

- To produce a more interesting plot, we represent the precondition as a new goal that aims to get the relevant information to rescue the hostage. The goal is referred to as *The hero finds an informant that provides the location of the hostage and the instructions to rescue him*.

Next, we focus on this new goal.

(5) GOAL: *The hero finds an informant that provides the location of the hostage and the instructions to rescue him* (description of its preconditions).

This goal has two preconditions (see Figure 4.2d):

- The first precondition requires that the character that will play the role of the informant be selected. In this way, the system can choose a different informant each time it generates a tale.
- With the purpose of incorporating an additional challenge to the plot, we decided that our hero needs to reward the informant to get the information. So we represent the second precondition as a new goal known as *The hero rewards the informant.*

Again, the decision to include these two preconditions is based on our knowledge about informants, and on our belief that inserting an additional challenge at this point in the plot helps to produce a more interesting story. Next, we focus on this further goal.

(6) GOAL: *The hero rewards the informant* (description of its precondition and plan).

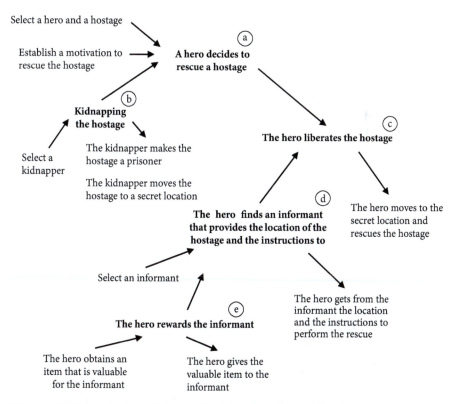

Figure 4.2 The goal-oriented plot graph of the story about abduction

This goal has one precondition, and its plan has one action to perform (see Figure 4.2e):

- The precondition requires that the hero obtains an item that is valuable to the informant. When we define the data structures, we will establish what makes an item valuable.
- The goal's plan consists of the hero giving that valuable item to the informant.

We return to the goal *The hero finds an informant that provides the location of the hostage and the instructions to rescue him*. Because we have defined all its preconditions, we can establish its plan.

(7) GOAL: *The hero finds an informant that provides the location of the hostage and the instructions to rescue him* (description of its plan).

- The plan indicates one action (see Figure 4.2d): the hero gets from the informant the location and instructions to perform the rescue.

We return to the goal *The hero liberates the hostage*. Because we have defined all its preconditions, we focus on its plan.

(8) GOAL: *The hero liberates the hostage* (description of its plan).

- The plan includes one action (see Figure 4.2c): the hero moves to the secret location and rescues the hostage.

With this, the GOP-graph is finished. When a goal includes another goal as part of its preconditions or its plan, we refer to the latter as a nested goal. A plot might have several nested goals. Figure 4.2 illustrates this case. The plan of the driving goal *The hero decides to rescue the hostage* includes the goal *The hero liberates the hostage*, which in turn has as a precondition, the goal *The hero finds an informant that provides the location of the hostage and the instructions to rescue him*, which itself has as a precondition, the goal *The hero rewards the informant*. Nested goals are useful for elaborating complex, and hopefully more interesting, plots.

The GOP-graph provides essential information for designing an automatic storyteller. In the rest of this chapter, we will employ the graph shown in Figure 4.2 to establish the features of the story-world and to define the data structures of our system.

4.3 Characteristics of the story-world

The GOP-graph in Figure 4.2 can be used to generate narratives about hostage taking in multiple settings. Thus, the first step is to choose a setting and define its features. For this example, we employ a fantastic world, which typically is inhabited by kings, princesses, wizards, and so on, who dwell in castles and enchanted places, among

Locations: castle, cursed cave, sorcerer's house, sacred lake, mother mountain, ancient tree, temple, market.

Characters: old and wise king, princess, lord of evil, sorcerer, soothsayer, priest, guardian of the lake, knight, old mysterious lady.

Gender: male, female, non-binary.

Objects: magic grass, foretelling bones, venerated book, magic sword.

Figure 4.3 Basic elements for the design of an automatic storyteller

others. Our plot requires at least four characters, to play the following roles: the hero, the hostage, the kidnapper, and the informant. For this exercise, we aim to develop a system that automatically selects which character plays each role (rather than the user choosing the protagonists, as in Chapter 3). Thus, we need a set of actors from which the program can pick one. We have defined nine: the old and wise king, the princess, the lord of evil, the sorcerer, the soothsayer, the priest, the guardian of the lake, the knight, and the old mysterious lady. You can include your own characters if you wish. They have a gender: male, female, non-binary. The story-world requires locations where the characters interact. For this exercise, we define the following: the castle, the cursed cave, the sorcerer's house, the sacred lake, the mother mountain, the ancient tree, the temple, and the market. Because the plot requires that the hero gives a valuable item to the informant (see Figure 4.2e), we need to incorporate objects in the story-world. Thus, we include the following articles: magic grass, foretelling bones, venerated book, and the magic sword. Figure 4.3 shows these basic elements.

Once we have established the features of the story-world, we focus on defining the attributes of the data structures that represent characters.

4.4 Defining the character-structures

Figure 4.4 shows the attributes of the character-structure, namely, how data are specified for each character in the story. The first three attributes are *Name, which identifies the character, *Gender, which specifies its gender, and *Location, which registers its location. Figure 4.3 shows the possible values that these three variables can take. In the following, we explain the rest of the character attributes and their possible values.

Our story includes kidnappers, heroes, and so on (see Figure 4.2). If we analyse other well-known fictional worlds, the contrast between the personalities of the protagonist and the antagonist is an important narrative element. Knowing that a villain is a bad person, or that the hero is a brave individual, helps the audience to make sense of their actions. Thus, we need to represent this feature in our system.

Character-structure

*Name: [old and wise king, princess, lord of evil, sorcerer, soothsayer, priest, guardian of the lake, knight, old mysterious lady]
*Gender: [male, female, non-binary].
*Location: [castle, cursed cave, sorcerer's house, sacred lake, mother mountain, ancient tree, temple, market]
*Emotion towards other character: [love, indifference, hate, none]
*Character who is the object of the emotion: [old and wise king, princess, lord of evil, sorcerer, soothsayer, priest, guardian of the lake, knight, old mysterious lady, none]
*Personality: [brave, friendly, mean, horrendous]
*Captivity-status: [free, imprisoned]
*Reliable source of information: [yes, no]
*Object that the character owns: [magic grass, foretelling bones, venerated book, magic sword, none]
*Location that the character learns: [castle, cursed cave, sorcerer's house, sacred lake, mother mountain, ancient tree, temple, market, none]
*Instructions to perform: [Any story-action, none]

Figure 4.4 Character-structure and all possible values for each attribute

The attribute *Personality registers the temperament of the actor; there are four possible values: *brave, friendly, mean, horrendous*. The designer of the system can define any number of types of personalities. In this way, it is possible to have a brave princess, horrendous lord of evil, and so on.

Our story requires that it be explained why the hero decides to rescue the hostage (see Figure 4.2a). Again, we rely on the work of experts to obtain answers. It is common to find action movies where the protagonist's spouse, son, or father is kidnapped. In these films, the strong emotional bond between the family members explains why one of them comes to help the others. Inspired by this, we include two attributes that represent emotional links between characters. The first is *Emotion towards other character* that indicates the kind of emotional bond that characters can have with each other. It has four possible values: *love, indifference, hate, none*. The value *none* indicates that the actor has not developed any emotion towards someone else. The second is *Character who is the object of the emotion* that indicates who is the recipient of the emotion. In this way, it is possible to represent that the princess loves the king, which would explain why she decides to help him.

Our story requires to find an informant (see Figure 4.2d). The attribute *Reliable source of information* specifies if a particular actor has special skills, for example magical powers, to offer information that is relevant for another individual. It has two possible values: *yes* or *no*. In this way, this attribute can be set to true for characters like the sorcerer or the soothsayer.

Our story requires that characters manipulate objects (see Figure 4.2e). The attribute *Objects that the character owns* records the things that this character possesses. Its possible values are *magic grass, foretelling bones, venerated book, magic sword, none*. Characters must be able to discover hidden locations. The attribute *Location that the character learns* registers those locations that this character learns from others. Its possible values are *castle, cursed cave, sorcerer's house, sacred lake, mother mountain, ancient tree, temple, market, none*. Similarly, characters must be able to obtain instructions to execute (see Figure 4.2d). Thus, *Instructions to perform* stores instructions (story-actions to perform) that this character learns from others.

Table 4.1 Attributes' default values for the nine character-structures

*Name	old and wise king	princess	lord of evil	sorcerer	soothsayer	priest	guardian of the lake	knight	old mysterious lady
*Location	castle	castle	cursed cave	sorcerer's house	mother mountain	temple	sacred lake	castle	ancient tree
*Gender	male	female	male	male	female	male	non-binary	male	female
*Emotion towards other character	none	none	none	none	none	none	none	none	none
*Character who is the object of the emotion	none	none	none	none	none	none	none	none	none
*Personality	friendly	brave	horrendous	mean	mean	friendly	horrendous	brave	friendly
*Captivity-status	free	free	free	free	free	free	free	free	free
*Reliable source of information	no	no	no	yes	yes	no	no	no	yes
*Object that the character owns	none	none	none	none	none	none	none	none	none
*Location that the character learns	none	none	none	none	none	none	none	none	none
*Instructions to perform	none	none	none	none	none	none	none	none	none

And, of course, characters might or might not be hostages. The attribute *Captivity-status* records this condition, and it has two possible values: *free* or *imprisoned*.

Figure 4.4 shows the character-structure and, inside square brackets, all the possible values that each attribute can have. Thus, complex human features like personality or emotional relationships are represented as attributes that the computer can work with.

For the example in this chapter, we have decided to predefine all characters; that is, the user cannot incorporate new actors. In this way, we, as designers, can assign default values to characters' attributes. For instance, we determine that, by default, the personality of the princess is brave, the king is friendly, and the lord of evil is horrendous; in a similar way, we decide that the personality of the sorcerer is mean and that he is a reliable source of information. We can also predefine the initial location of all characters, and so on. Thus, when the system starts, nine character-structures are built, one for each character, and the default values of all attributes are assigned (see Table 4.1). We will employ these features to make some actions happen. Assigning default values to attributes makes it possible to build consistent character profiles.

The next step is to define the structures that represent the system's character-goals.

4.5 Defining the goal-structures

The GOP-graph (see Figure 4.2) shows that our plot has five goals:

(a) A hero decides to rescue a hostage.
(b) Kidnapping the hostage.
(c) The hero liberates the hostage.
(d) The hero finds an informant that provides the location of the hostage and the instructions to rescue him.
(e) The hero rewards the informant.

The graph also shows how goals are linked. The following describes how we use the GOP-graph as a guide to establish the attributes of the goal-structures.

4.5.1 Goal-structure: a hero decides to rescue a hostage

Figure 4.2a provides general information about this goal: it has three preconditions, and the plan includes one goal. Based on this information, the next step is to define its attributes. Thus, this goal-structure is organized as follows (see Figure 4.5):

- The goal involves two participants: *Character-Hero* and *Character-Hostage*.
- The first precondition indicates that the system must instantiate the variables *Character-Hero* and *Character-Hostage*. For the former, the program looks for all those character-structures whose attribute *Personality* is set to *brave*

and then chooses one of them at random. The purpose of this procedure is to find an actor whose personality fits the role of a hero. From now onwards, any reference to *Character-Hero is interpreted as a reference to this associated structure. So, if the princess's character-structure is randomly picked, then she becomes the hero in the story. The system follows the same procedure to instantiate the variable *Character-Hostage. This time the program looks for all those character-structures whose attribute *Personality is set to *friendly* and then chooses one of them at random. This case illustrates how the designer might include as part of a goal-structure a set of commands that the system must perform each time the goal becomes active. We employ in the goal-structure the string 'SYS' to signal this type of directions. Similarly, this case illustrates the use of a randomizing mechanism to avoid deterministic outputs, and the use of relevant characters' attribute to constrain the selection of characters. Because the hero and the hostage are chosen at random, each time the program generates a new tale the protagonists might be different.

- The second precondition executes the action *Character-Hero loves *Character-Hostage. Its aim is to establish an emotional bond between them, which later explains why the hero helps the hostage. Thus, with the help of the GOP-graph, while we are defining the goals, at the same time we area also figuring out the story-actions that our plot needs. Later, we will outline their structures.
- The third precondition activates the goal *Kidnapping the hostage*.
- Finally, the plan triggers a new goal named *The hero liberates the hostage*.

Goal-structure

Name of the goal: A hero decides to rescue a hostage
*Character-Hero
*Character-Hostage

Preconditions:
- SYS: To instantiate *Character-Hero find all characters whose attribute *Personality is set to *brave* and choose one at random. To instantiate *Character-Hostage find all characters whose attribute *Personality is set to *friendly* and choose one at random.
- ACT: *Character-Hero **loves** *Character-Hostage.
- GOAL: Kidnapping the hostage.

Plan:
-GOAL: The hero liberates the hostage.

Figure 4.5 Representation of the goal *A hero decides to rescue a hostage*

4.5.2 Goal-structure: kidnapping the hostage

Based on Figure 4.2b, this goal-structure is organized as follows (see Figure 4.6):

- The goal involves two participants: *Character-Hostage and *Character-Kidnapper.

Goal-structure

Name of the goal: Kidnapping the hostage
*Character-Hostage
*Character-Kidnapper

Preconditions:
- SYS: Instantiate the *Character-Kidnapper based on the following procedure:
     (Find all characters whose attribute *Personality is set to *horrendous* and choose one at random) OR
     (Find all characters whose attributes *Emotion towards other character* is set to *hate*, *Character
     who is the object of emotion is equal to *Character-Hostage, and choose one of them at random).

Plan:
ACT: *Character-Kidnapper **makes prisoner** *Character-Hostage.
ACT: *Character-Kidnapper **takes to the cursed cave** *Character-Hostage.

Figure 4.6 Representation of the goal-structure *Kidnapping the hostage*

- The precondition indicates that the system must instantiate *Character-Kidnapper* by performing at random either of the following two procedures:

  (i) Look for all those characters whose attribute *Personality* is set to *horrendous* and then choose one of them at random. The purpose of this procedure is to find an actor whose personality fits the role of a captor.

  (ii) Look for all those characters whose attributes *Emotion towards other character* is set to hate, *Character who is the object of emotion* is equal to *Character-Hostage*, and then choose one of them at random. The purpose of this procedure is to find an actor that the captor hates.

- The plan consists of performing two actions. The first, *Character-Kidnapper makes prisoner* *Character-Hostage*, makes the kidnapper imprison the hostage. The second, *Character-Kidnapper takes to the cursed cave* *Character-Hostage*, locates both characters inside the cursed cave.

4.5.3 Goal-structure: the hero liberates the hostage

Based on Figure 4.2c, this goal-structure is organized as follows (see Figure 4.7):

- The goal involves two participants: *Character-Hero* and *Character-Hostage*.
- The precondition triggers a goal named *The hero finds an informant that provides the location of the hostage and the instructions to rescue him*.
- The plan executes two actions. The first, *Character-Hero moves to the location that the character learns*, places the actor in the location that is recorded in its attribute *Location that the character learns*. The second, *Character-Hero executes the instructions to perform*, performs the action that is recorded in the actor's attribute *Instructions to perform*.

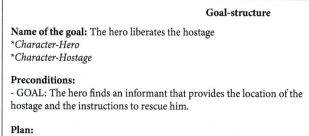

> **Goal-structure**
>
> **Name of the goal:** The hero liberates the hostage
> *Character-Hero*
> *Character-Hostage*
>
> **Preconditions:**
> - GOAL: The hero finds an informant that provides the location of the hostage and the instructions to rescue him.
>
> **Plan:**
> ACT: *Character-Hero* **moves to the location that the character learns.**
> ACT: *Character-Hero* **executes the instructions to perform**

Figure 4.7 Representation of the goal-structure *The hero liberates the hostage*

4.5.4 Goal-structure: the hero finds an informant that provides the location of the hostage and the instructions to rescue him/her

Based on Figure 4.2d, this goal-structure is organized as follows (see Figure 4.8):

- The goal involves two participants: *Character-Hero* and *Character-Informant*.
- The first precondition is designed to select an informant. Thus, the system looks for all character-structures whose attribute *Reliable source of information* is set to true, choose one at random, and then use it to instantiate the variable *Character-Informant*.

> **Goal-structure**
>
> **Name of the goal:** The hero finds an informant that provides the location of the hostage and the instructions to rescue him
> *Character-Hero*
> *Character-Informant*
>
> **Preconditions:**
> - SYS: Find all characters whose attribute *Reliable source of information is set to true*, choose one at random, and use it to instantiate *Character-Informant*.
> - ACT: *Character-Hero* **moves to the same location as** *Character-Informant*.
> - If *Character-Informant*'s attribute *Personality* is set to *mean* or to *horrendous*, then
>   GOAL: The hero rewards the informant
>
> **Plan:**
> - ACT: *Character-Hero* **gets the location of the hostage from** *Character-Informant*.
> - ACT: *Character-Hero* **gets instructions to rescue the hostage from** *Character-Informant*.

Figure 4.8 Representation of the goal-structure *The hero finds an informant that provides the location of the hostage and the instructions to rescue him*

- The second precondition executes the action *Character-Hero moves to the same location as *Character-Informant. So the hero is positioned at the informant's location.
- The third precondition is designed to produce some variety in the stories. Its aim is to employ the characters' personality to shape the scene. So if the informant is mean or horrendous, he will demand a reward as a condition to reveal the required information. Otherwise, he will willingly give the information to the hero. Although this situation is not explicitly stated in the GOP-graph, we believe it contributes to developing interesting plots.
- The plan has two actions: the first, *Character-Hero gets the location of the hostage from *Character-Informant, and the second, *Character-Hero gets instructions to rescue the hostage from *Character-Informant. In this way, the hero learns where the hostage is situated and how to rescue him.

4.5.5 Goal-structure: the hero rewards the informant

Based on Figure 4.2e, this goal-structure is organized as follows (see Figure 4.9):

- The goal involves two participants: *Character-Hero and *Character-Informant.
- There is one precondition. Its purpose is to determine which object will be employed as a valuable item in the tale. Because we do not know in advance who the informant is, we associate to each actor a valuable object. So if the informant is the sorcerer, then he will be fascinated by the magic grass; if the informant is the soothsayer, then he will be fascinated by the foretelling bones; and so on. In this way, the informant asks as a reward an object that is valuable to him.

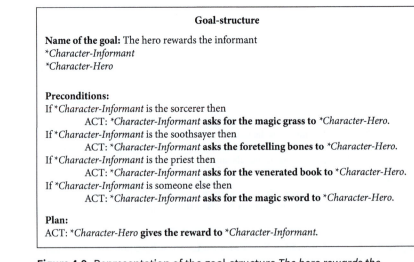

Goal-structure

Name of the goal: The hero rewards the informant
*Character-Informant
*Character-Hero

Preconditions:
If *Character-Informant is the sorcerer then
      ACT: *Character-Informant **asks for the magic grass to** *Character-Hero.
If *Character-Informant is the soothsayer then
      ACT: *Character-Informant **asks the foretelling bones to** *Character-Hero.
If *Character-Informant is the priest then
      ACT: *Character-Informant **asks for the venerated book to** *Character-Hero.
If *Character-Informant is someone else then
      ACT: *Character-Informant **asks for the magic sword to** *Character-Hero.

Plan:
ACT: *Character-Hero **gives the reward to** *Character-Informant.

Figure 4.9 Representation of the goal-structure *The hero rewards the informant*

For reason of space, in Section 4.6 we will only provide details of the action *asks for the magic grass to*. An advanced version of our program should include story-actions that cover all possibilities.

- The plan has one action, *Character-Hero gives the reward to *Character-Informant*.

Every goal-structure in this section includes at least one action to be performed, which is indicated by the string 'ACT'. As we explained in Chapter 3, every time the system finds the 'ACT' directive, it looks for the corresponding action-structure to execute the deed. Thus, the last step is to define the attributes of story-action structures.

4.6 Defining the story-action structures

The main purpose of story-actions is to modify the story-world. In this example, the story-world is represented by the characters' attributes, for example, their location, the objects they own, if they are free or imprisoned. Thus, performing a story-action results in modifying at least one character's attribute. In the following, we describe each story-action structure.

(a) Story-action structure: *Loves*. Its aim is to establish an emotional bond between two characters. Figure 4.10a shows its organization:
  - The deed includes two participants: *Character-that-loves* and *Character-that-is-loved*. The names of the variables illustrate the role of the characters in the action. During the generation of a tale, the system employs a mechanism to keep track of the different variables representing actors, and how they relate to each other. For instance, when the action *Character-Hero loves *Character-Hostage* is performed (see the preconditions in Figure 4.5), the system knows that the hero is the character who loves, and the hostage is the character who is loved.
  - There are no preconditions.
  - The post-condition modifies the story-world as follows: the *Character-that-loves*'s attributes *Emotion towards other character* is set to *Love*, and *Character who is the object of the emotion* is set to the same value as *Character-that-is-loved*.
  - In this chapter, we have added a new element to the story-action structures: templates (see Chapter 2). Templates comprise a text that describes how characters perform an action; these descriptions are printed as part of the output. For this action, we have defined two sentences: 'The [*Character-that-loves] loved the [*Character-that-is-loved]' and 'The [*Character-that-loves] had a strong affection for the [*Character-that-is-loved].' Once the characters in the action have been instantiated, the system substitutes the variables in the templates with the correct values. For consistency, all the

texts we employ are in the past tense. You can include as many templates as you require. During the generation of a story, the program chooses one of them at random.

(b) Story-action structure: *Makes prisoner*. Its aim is to make a character a hostage. Figure 4.10b shows its organization:

- The deed includes two participants: *Character-Kidnapper* and *Character-Hostage*.
- There is one precondition: the location of both characters must be different. Its purpose is to be sure that the kidnapper must go and look for the victim. In this way, the protagonists move around the story-world.
- The post-condition modifies the story-world as follows: the *Character-Kidnapper*'s attribute *Location* is set to the value of the *Character-Hostage*'s attribute *Location*. That is how the system moves one character to a different place. Then the *Character-Hostage*'s attribute *Captivity-status* is set to *imprisoned*.
- There is one template: 'One morning, the [*Character-Kidnapper*] went to the [*Character-Hostage* / *Location*] and abducted the [*Character-Hostage*].' The second variable in this template includes information about the current state of the story-world, in this case, the location of the hostage.

(c) Story-action structure: *Takes to the cursed cave*. Its aim is to move the actors to the cursed cave. Figure 4.10c shows its organization:

- The deed includes two participants: *Character-Kidnapper* and *Character-Hostage*.
- There is no precondition.
- The post-condition modifies the story-world as follows: the *Character-Kidnapper*'s attribute *Location* is set to *cursed cave*. The *Character-Hostage*'s attribute *Location* is set to *cursed cave*.
- There is one template: 'The [*Character-Kidnapper*] took the [*Character-Hostage*] to the cursed cave.'

(d) Story-action structure: *Moves to the location that the character learns*. Its aim is to move a character from its current location to the location that it learned from other character. Figure 4.10d shows its organization:

- The deed includes one participant: *Character-that-changes-position*.
- There is no precondition.
- The post-condition modifies the story-world as follows: the *Character-that-changes-position*'s attribute *Location* is set to the value of its attribute *Location that the character learns*.
- There is one template: 'Wasting no time, the [*Character-that-changes-position*] went to the [*Character-that-changes-position* / *Location that the character learns*].' Again, this text illustrates the use of a character's attribute.

(e) Story-action structure: *Executes the instructions to perform*. Its aim is to perform the action that the character learnt from another actor. Figure 4.10e shows its organization:

- The deed includes one participant: *Character-Executer*.
- There is no precondition.

- The post-condition modifies the story-world as follows: the system performs the action in the *Character-Executer*'s attributes *Instructions to perform*.
- There are no templates.

(a) Story-action structure

Name of the action: Loves
*Character-that-loves
*Characters-that-is-loved

Preconditions: None.
Post-conditions: The *Character-that-loves*'s attribute *Emotion towards other character* is set to Love and *Character who is the object of the emotion* is set to the same value as *Character-that-is-loved*.

TEMPLATES:
- The [*Character-that-loves] loved the [*Character-that-is-loved].
- The [*Character-that-loves] had a strong affection for the [*Character-that-is-loved].

(b) Story-action structure

Name of the action: Makes prisoner
*Character-Kidnapper
*Character-Hostage

Preconditions: The location of both characters is different.
Post-conditions: The *Character-Kidnapper*'s attribute *Location* is set to the value of the *Character-hostage*'s attribute *Location*.
The *Character-hostage*'s attribute *Captivity-status* is set to imprisoned.

TEMPLATES:
- One morning, the [*Character-Kidnapper] went to the [*Character-Hostage/*Location] and abducted the [*Character-hostage].

(c) Story-action structure

Name of the action: Takes to the cursed cave
*Character-Kidnapper
*Character-Hostage

Preconditions: None.
Post-conditions: The *Character-Kidnapper*'s attribute *Location* is set to cursed cave.
The *Character-Hostage*'s attribute *Location* is set to cursed cave.

TEMPLATES:
The [*Character-Kidnapper] took the [*Character-Hostage] to the cursed cave

---

(d) Story-action structure

Name of the action: Moves to the location that the character learns
*Character-that-changes-position

Preconditions: None.
Post-conditions: The *Character-that-changes-position*'s attribute *Location* is set to the value of its attribute *Location that the character learns*.

TEMPLATES:
- Wasting no time, the [*Character-that-changes-position] went to the [*Character-that-changes-position/*Location that the character learns].

(e) Story-action structure

Name of the action: Executes the instructions to perform
*Character-Executer

Preconditions: None.
Post-conditions: Perform the action in the *Character-Executer*'s attributes *Instructions to perform*.

TEMPLATES: none.

Figure 4.10 Structures of the story-actions (a) *Loves*, (b) *Makes prisoner*, (c) *Takes to the cursed cave*, (d) *Moves to the location that the character learns*, and (e) *Executes the instructions to perform*

Executes the instructions to perform is a special action because it employs a second action, the one recorded in the attribute *Instructions to perform*, to modify the story-world. That is why it does not require templates.

(f) Story-action structure: *Moves to the same location as*. Its aim is to move a character to the location of a second character. Figure 4.11f shows its organization:

- The deed includes two participants: *Character-that-moves* and *Character-Static*.
- There is no precondition.
- The post-condition modifies the story-world as follows: The *Character-that-moves*'s attribute *Location* is set to the same value as the *Character-Static*'s attribute *Location*.
- There is one template: 'The [*Character-that-moves*] went to talk to the [*Character-Static*] to ask for help.'

(g) Story-action structure: *Gets the location of the hostage from*. Its aim is to obtain from the informant the place where the hostage is kept captive. Figure 4.11g shows its organization:
- The deed includes three participants: *Character-Hero*, *Character-Informant*, and *Character-Hostage*.
- There is no precondition.
- The post-condition modifies the story-world as follows: The *Character-Hero*'s attribute *Location that the character learns* is set to the same value as the *Character-Hostage*'s attribute *Location*.
- There is one template: 'The [*Character-Informant*] revealed to the [*Character-Hero*] the [*Character-Hostage*]'s location.'

(h) Story-action structure: *Gets instructions to rescue the hostage from*. Its aim is that the hero gets a one-action instruction from the informant to liberate the hostage. Figure 4.11h shows its organization:
- The deed includes two participants: *Character-Hero* and *Character-Informant*.
- There is one precondition: The *Character-Hostage*'s attribute *Captivity-status* is equal to *imprisoned*. Otherwise, the action makes no sense.
- The post-condition modifies the story-world as follows: The *Character-Hero*'s attribute *Instructions to perform* is set to *Character-Hero performs the rescue plan The spell*.
- There is one template: 'The [*Character-Informant*] gave to the [*Character-Hero*] a spell to put the [*Character-Kidnapper*] to sleep.'

In this action, the hero always receives the same instruction to save the victim. In a more advanced version of the system, we could incorporate different rescue strategies. In this way, our narratives would be more diverse.

(i) Story-action structure: *Performs the rescue plan The spell*. Its aim is to cast a spell on the kidnapper to rescue the hostage. Figure 4.11i shows its organization:
- The deed includes three participants: *Character-Hero*, *Character-Kidnapper*, and *Character-Hostage*.
- There is one precondition: The *Character-Hostage*'s attribute *Captivity-status* is equal to *imprisoned*. Otherwise, the action makes no sense.

- The post-condition modifies the story-world as follows: The *Character-Hostage*'s attribute *Captivity-status* is set to free.
- The template is arranged as follows: 'The [*Character-Hero*] invoked the spell and the [*Character-Kidnapper*] fell into a deep sleep. The [*Character-Hero*] released the [*Character-Hostage*].'

(f) **Story-action structure**

Name of the action: Moves to the same location as
Character-that-moves
Character-Static

Preconditions: None.
Post-conditions: The *Character-that-moves*'s attribute *Location* is set to the same value as the *Character-Static*'s attribute *Location*.

TEMPLATES:
The [*Character-Hero*] went to talk to the [*Character-Informant*] to ask for help

---

(g) **Story-action structure**

Name of the action: Gets the location of the hostage from
Character-Hero
Character-Informant
Character-Hostage

Preconditions: none.
Post-conditions: The *Character-Hero*'s attribute *Location that the character learns* is set to the same value as the *Character-Hostage*'s attribute *Location*.

TEMPLATES:
- The [*Character-Informant*] revealed to the [*Character-Hero*] the [*Character-Hostage*]'s location.

(h) **Story-action structure**

Name of the action: Gets instructions to rescue the hostage from
Character-Hero
Character-Informant

Preconditions: The *Character-Hostage*'s attribute *Captivity-status* is equal to imprisoned.
Post-conditions: The *Character-Hero*'s attribute *Instructions to perform* is set to *Character-Hero performs the rescue plan The spell*.

TEMPLATES:
The [*Character-Informant*] gave to the [*Character-Hero*] a spell to put the [*Character-Kidnapper*] to sleep.

(i) **Story-action structure**

Name of the action: Performs the rescue plan The spell
Character-Hero
Character-Kidnapper
Character-Hostage

Preconditions: The *Character-Hostage*'s attribute *Captivity-status* is equal to imprisoned.
Post-conditions: The *Character-Hostage*'s attribute *Captivity-status* is set to free.

TEMPLATES:
- The [*Character-Hero*] invoked the spell and the [*Character-Kidnapper*] fell into a deep sleep. The [*Character-Hero*] released the [*Character-Hostage*].

(j) **Story-action structure**

Name of the action: Asks for the magic grass
Character-Hero
Character-Informant

Preconditions: The *Character-Informant*'s attribute *Object that the character owns* is different to magic grass.

Post-conditions: The *Character-Hero*'s attribute *Object that the character owns* is set to magic grass.

TEMPLATES:
- The [*Character-Informant*] agreed to provide the information in exchange for a cluster of magic grass that only grows in the sacred lake. The [*Character-Hero*] went in search of the plant but the guardian of the lake barred [her/him] from entering. The [*Character-Hero*] sang an ancestral song of supplication and the guardian of the lake allowed [her/him] to enter. The [*Character-Hero*] took a bunch of magic grass. The [*Character-Hero*] returned to the [*Character-Informant*].

(k) **Story-action structure**

Name of the action: Gives the reward to
Character-Hero
Character-Informant

Preconditions: None
Post-conditions: The *Character-Informant*'s attribute *Object that the character owns* is set to the value of the *Character-Hero*'s attribute *Object that the character owns*. Then, the *Character-Hero*'s attribute *Object that the character owns* is set to none.

TEMPLATES:
- The [*Character-Hero*] handed to the [*Character-Informant*] the [*Character-Hero/*Object that the character owns*].

Figure 4.11 Structures of the story actions (f) *Moves to the same location as*, (g) *Gets the location of the hostage from*, (h) *Gets instructions to rescue the hostage from*, (i) *Performs the rescue plan The spell*, (j) *Asks for the magic grass to*, and (k) *Gives the reward to*

Thus, the system releases a captive by setting its *Captivity-Status attribute to the value of *free*. However, it employs templates to describe this simple action as a complex scenario involving sleeping spells.

(j) Story-action structure: *Asks for the magic grass*. Its aim is for the informant to demand the magic grass. Figure 4.11j shows its organization:
- The deed includes two participants: *Character-Hero* and *Character-Informant*.
- There is one precondition: The *Character-Informant*'s attribute *Object that the character owns* must be different to *magic grass*. Otherwise, the action makes no sense.
- The post-condition modifies the story-world as follows: The *Character-Hero*'s attribute *Object that the character owns* is set to *magic grass*.
- The template comprises the following text: 'The [*Character-Informant] agreed to provide the information in exchange for a cluster of magic grass that only grows in the sacred lake. The [*Character-Hero] went in search of the plant but the guardian of the lake barred [her/him/them] from entering. The [*Character-Hero] sang an ancestral song of supplication and the guardian of the lake allowed [her/him/them] to enter. The [*Character-Hero] took a bunch of magic grass. The [*Character-Hero] returned to the [*Character-Informant].'

This case illustrates how templates can be used to incorporate an entire passage into a story. We will discuss later its advantages and limitations. Also, this case shows the use of pronouns. Depending on the value of the character's attribute *Gender*, the program chooses the right pronoun.

(k) Story-action structure: *Gives the reward to*. Its aim is to deliver the valuable item to the informant. Figure 4.11k shows its organization:
- The deed includes two participants: *Character-Hero, *Character-Informant*.
- There is no precondition.
- The post-condition modifies the story-world as follows: The *Character-Informant*'s attribute *Object that the character owns* is set to the value of the *Character-Hero*'s attribute *Object that the character owns*. Then the *Character-Hero*'s attribute *Object that the character owns* is set to *none*.
- There is one template: 'The [*Character-Hero] handed to the [*Character-Informant] the [*Character-Hero / *Object that the character owns].'

4.7 Final remarks about our methodology

The methodology that we have introduced describes the main steps necessary to develop an automatic narrative generator. Throughout the chapter we have employed a random selection of options as a tool to avoid developing a deterministic system. The rich story-world makes it possible to combine characters, objects, personalities, and so on, to create diverse narrative contexts. The use of rules is useful for shaping scenes based on the current value of characters' attributes. Nested goals permit building elaborated plots. You as a designer can choose where to include new goals. As we explained earlier, as a rule of thumb important circumstances must be represented as goals to introduce new challenges to the protagonists at critical moments in the plot.

Templates help to transform the data-structures that comprise a story into text. They can be used to produce a range of text, from austere descriptions to elaborated passages. For instance, an action that moves a character to a cabin might be expressed as 'The princes went to the cabin' or as 'That sunny morning the princess decided to take the dangerous path to get to the cabin as soon as possible'. The former is a basic description that can be employed in multiple contexts, for example, stories set either during the day or the night. The latter is a richer description, but its use is constrained to scenes happening during the morning in places close to dangerous paths. Thus, the longer the text, the more difficult it is to use it in various narrative contexts.

The system presented in this chapter can be expanded in multiple ways. For instance, how should we handle those situations where the preconditions of an action are not satisfied? One option is to decide that the story cannot be completed and then stop the generation process. A second option is to develop routines that allow the program to solve the problem. Basically, these routines search among the set of available story-actions for those whose post conditions satisfy the unfulfilled preconditions. Let us illustrate this case. Throughout the chapter we have made sure that those actors that interact with each other are in the same place. This is a common-sense precondition that applies to all deeds. However, a more elaborated GOP-graph might include circumstances that produce a context in which two characters located at different places attempt to participate in the same deed. Thus, the system requires a routine that detects this situation and then includes in the story an action that situates both characters in the same position before the problem arises. For this solution to work, it is necessary to have a set of story-actions whose post-conditions move a character to any spot in the story-world.

In Chapter 5 we will put into practice all the concepts we have learned.

5

Generating a narrative step by step

5.1 A story about a kidnapped king

In this chapter we show how a computer program generates a narrative using the structures that we designed in Chapter 4. When the program starts, *A hero decides to rescue a hostage* is set as the driving goal (see Figure 4.2; for the sake of clarity, in this section we make constant references to figures and tables introduced in Chapter 4). Next, the system assigns the default settings to each of its nine available characters. Thus, the lord of evil's *Personality* is set to *horrendous*, the sorcerer's *Personality* is set to *mean,* and his attribute *Reliable source of information* is set to *yes*, and so on. The location of each character is set to their natural position: for example, the old and wise king is located at the castle, the sorcerer is located at his home, the guardian of the lake is located at the sacred lake. See Table 4.1 for a complete list of settings. Next, the program focuses on the driving goal-structure *A hero decides to rescue a hostage* (see Figure 4.5). The first precondition triggers a routine to instantiate the main actors. Suppose that, at random, *Character-Hero* is set to *princess* and *Character-Hostage* is set to *old and wise king*. Figure 5.1 shows the instantiated structure.

Once the main characters in the tale are chosen, the system prints a sentence that works as an introduction to the reader. Employing templates, these texts describe some of the characters' attributes (e.g. locations, personality). For instance, 'The [*Character] was at the [*Character/*Location]' or 'The [*Character] was a [*Character/*Personality] person'. Those sentences are chosen at random from a set of options defined in advance by the designer of the system. Thus, our example begins as follows:

The old and wise king was at the castle. The princess was a brave person.

The following goal's precondition requires executing the story-action where the princess loves the king (see Figures 5.1 and 4.10a). As a result of performing this action, the system sets the princess's attribute *Emotion towards other character* to *Love* and *Character who is the object of the emotion* to *old and wise king*. Next, it chooses at random one of the templates available to produce the output (see Figure 4.10a). Each time a story-action is performed by a character we show the text of the story in progress that has been produced up to that point; sentences in bold represent the last actions performed. Thus, the system inserts this new deed into the output:

An Introduction to Narrative Generators. Rafael Pérez y Pérez and Mike Sharples, Oxford University Press.
© Rafael Pérez y Pérez and Mike Sharples (2023). DOI: 10.1093/oso/9780198876601.003.0005

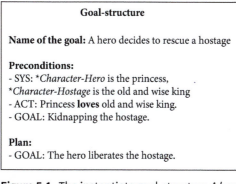

Goal-structure

Name of the goal: A hero decides to rescue a hostage

Preconditions:
- SYS: *Character-Hero* is the princess,
Character-Hostage is the old and wise king
- ACT: Princess **loves** old and wise king.
- GOAL: Kidnapping the hostage.

Plan:
- GOAL: The hero liberates the hostage.

Figure 5.1 The instantiate goal-structure *A hero decides to rescue a hostage*

The old and wise king was at the castle. The princess was a brave person. **The princess loved the old and wise king.**

The next precondition in Figure 5.1 triggers the goal *Kidnapping the hostage*. So the current goal is put on hold, and the system focuses on the new goal.

5.1.1 Processing the goal *Kidnapping the hostage*

The first precondition instructs the system to instantiate *Character-Kidnapper* (see Figure 4.6). The kidnapper must either have a horrendous personality or hate the king. Only the lord of evil and the guardian of the lake satisfy these requirements (see Table 4.1). The system chooses at random the former as the kidnapper. Figure 5.2 shows the instantiated structure.

Next, the system executes the goal's plan. First, it performs the action where the lord of evil makes the king a prisoner. So, the lord of evil moves to the same place as the king, that is, the lord of evil's attribute *Location* is set to the value of the king's attribute *Location*. Then, the system sets the king's *Captivity-status* to

---

Goal-structure

Name of the goal: Kidnapping the hostage
Character-Hostage is the old and wise king

Preconditions:
- SYS: *Character-Kidnapper* is the lord of evil.

Plan:
- ACT: Lord of evil **makes prisoner** old and wise king.
- ACT: Lord of evil **takes to the cursed cave** old and wise king.

Figure 5.2 Representation of the instantiated goal-structure *Kidnapping the hostage*

imprisoned (see Figure 4.10b). In the second action in the plan, the lord of evil takes the king to the cursed cave; so, for both characters *Location* is set to *cursed cave* (see Figure 4.10c). These actions are added to the output by printing their corresponding templates:

> *... The princess loved the old and wise king.* **One morning, the lord of evil went to the castle and abducted the old and wise king. The lord of evil took the old and wise king to the cursed cave.**

Because all the actions from the plan have been performed, the goal of having the king kidnapped has been achieved. The program gets back to the driving goal of rescuing the king, as shown in Figure 5.1. Because all its preconditions are fulfilled, the plan is put into action; thus, the goal *The hero liberates the hostage* is triggered (see Figure 4.7).

5.1.2 Processing the goal *The hero liberates the hostage*

All characters participating in this goal have been already defined. So the system instantiates this goal-structure (see Figure 5.3). The precondition triggers the goal *The hero finds an informant that provides the location of the hostage and the instructions to rescue him*, so, the current goal is put on hold.

Goal-structure

Name of the goal: The hero liberates the hostage
Character-Hero is the princess
Character-Hostage is the old and wise king

Preconditions:
- GOAL: The hero finds an informant that provides the location of the hostage and the instructions to rescue him.

Plan:
- ACT: Princess **moves to the position learned.**
- ACT: Princess **executes the plan learned**

Figure 5.3 Representation of the instantiated goal-structure *The hero liberates the hostage*

5.1.3 Processing the goal *The hero finds an informant that provides the location of the hostage and the instructions to rescue him*

The first precondition instructs the system to instantiate *Character-Informant* with any character whose attribute *Reliable source of information* is set to *true*

Goal-structure

Name of the goal: The hero finds an informant that provides the location of the hostage and the instructions to rescue him
Character-Hero is the princess

Preconditions:
- SYS: *Character-Informant* is the sorcerer
- ACT: Princess **moves to the same location as** sorcerer.
- If sorcerer's attribute *Personality* is set to *mean* or to *horrendous*, then
        GOAL: The hero rewards the informant

Plan:
- ACT: Princess **gets the location of the hostage from** sorcerer.
- ACT: Princess **gets instructions to rescue the hostage from** sorcerer.

Figure 5.4 Representation of the instantiated goal-structure *The hero finds an informant that provides the location of the hostage and the instructions to rescue him*

(see Figure 4.8). So the system chooses at random the sorcerer to play the role of informant. Figure 5.4 shows the resulting instantiated structure.

Next, the action where the princess moves to the same location as the sorcerer is performed. That is, the princess's attribute *Location* is set to the same value as the sorcerer's attribute *Location* (see Figure 4.11j). The new action is added to the tale:

> . . . *The lord of evil took the old and wise king to the cursed cave.* **The princess went to talk to the sorcerer to ask for help.**

The second precondition involves a rule. If the condition is not satisfied, the rule is ignored. Because the sorcerer has a mean personality (see Table 4.1), the goal *The hero rewards the informant* is triggered and the active goal in Figure 5.4 is put on hold.

5.1.4 Processing the goal *The hero rewards the informant*

All characters participating in this goal have been already defined. The preconditions include four rules to choose the most suitable reward for the informant (see Figure 4.9). Because only the first of those rules can be satisfied, all the preconditions can be simplified to the performing of the action where the sorcerer asks for the magic grass to the princess (see Figure 4.11j). Figure 5.5 shows the resulting instantiated structure.

As a result of performing the action, the princess's attribute *Objects that the character owns* is set to *magic grass*, and the text of the deed is added to the tale:

> . . . *The princess went to talk to the sorcerer to ask for help.* **The sorcerer agreed to provide the information in exchange for a cluster of magic grass that only grows**

in the sacred lake. The princess went in search of the plant but the guardian of the lake barred her from entering. The princess sang an ancestral song of supplication and the guardian of the lake allowed her to enter. The princess took a bunch of magic grass. The princess returned to the sorcerer.

Goal-structure

Name of the goal: The hero rewards the informant
*Character-Informant is the sorcerer
*Character-Hero is the princess

Preconditions:
- ACT: Sorcerer **asks for the magic grass to** princess.

Plan:
- ACT: Princess **gives the reward to** sorcerer.

Figure 5.5 Representation of the instantiated goal-structure *The hero rewards the informant*

Next, because the precondition is already satisfied, the plan is put into action. The system performs the action where the princess gives the reward to the sorcerer (see Figure 4.11k). As a result, the attributes **Objects that the character owns* for the princess and the sorcerer are updated; the system inserts this action into the narrative:

> *. . . The princess returned to the sorcerer.* **The princess handed to the sorcerer the magic grass.**

The plan is completed and, therefore, this goal has been achieved.

5.1.5 Going backwards

The system goes back to the previous goal *The hero finds an informant that provides the location of the hostage and the instructions to rescue him* (see Figure 5.4). Because all its preconditions are satisfied, the system performs the actions in the plan: *Princess gets the location of the hostage from Sorcerer* and *Princess gets instructions to rescue the hostage from Sorcerer* (see Figures 4.11g and 4.11h). As a result, the princess's attribute **Locations that the character learns* is updated with the current location of the king; similarly, the Princess' attribute **Instructions to perform* is set to *Princess performs the rescue plan The spell*. The system inserts these two actions in the tale:

> *. . . The princess handed to the sorcerer the magic grass.* **The sorcerer revealed to the princess the old and wise king's location. The sorcerer gave to the princess a spell to put the lord of evil to sleep.**

The plan is completed and, therefore, this goal has been achieved. The system goes back to the previous goal *The hero liberates the hostage* (see Figure 5.3). Because all preconditions are satisfied, the system performs the two actions that make up the plan: *Princess moves to the position learned* (i.e. to the cursed cave; see Figure 4.10d) and *Princess executes the plan learned* (see Figure 4.10e); i.e. the princess performs the action stored in her attribute **Instruction to perform*. In this case, this action is *Princess performs the rescue plan The spell* (see Figure 4.11i). These actions are inserted into the tale.

> *. . . The sorcerer gave to the princess a spell to put the lord of evil to sleep.* **Wasting no time, the princess went to the cursed cave. The princess invoked the spell and the lord of evil fell into a deep sleep. The princess released the king.**

The plan is completed and, therefore, this goal has been achieved. Thus, the driving goal has been successfully achieved too (see Figure 5.1). The story is finished by inserting the words 'The end'.

> *The old and wise king was at the castle. The princess was a brave person. The princess loved the old and wise king. One morning, the lord of evil went to the castle and abducted the old and wise king. The lord of evil took the old and wise king to the cursed cave. The princess went to talk to the sorcerer to ask for help. The sorcerer agreed to provide the information in exchange for a cluster of magic grass that only grows in the sacred lake. The princess went in search of the plant but the guardian of the lake barred her from entering. The princess sang an ancestral song of supplication and the guardian of the lake allowed her to enter. The princess took a bunch of magic grass. The princess returned to the sorcerer. The princess handed to the sorcerer the magic grass. The sorcerer revealed to the princess the old and wise king's location. The sorcerer gave to the princess a spell to put the lord of evil to sleep. Wasting no time, the princess went to the cursed cave. The princess invoked the spell and the lord of evil fell into a deep sleep. The princess released the old and wise king. The end.*

5.1.6 Improving the text

The use of routines to manipulate text are useful for producing a better-quality output. The following describes three examples.

(1) Employing equivalent words. The sequence of words 'old and wise king' is repeated several times in the tale. Because we know in advance the names of all the characters, it is possible to create a set of equivalent words that can substitute some instances of the repeated sequence, such as 'monarch', 'him', 'sovereign'.

(2) Joining two sentences. The word 'and' can be used to join two consecutive sentences when they share the same subject. For instance, 'The princess returned to the sorcerer. The princess handed to the sorcerer the magic grass' can be rewritten as 'The princess returned to the sorcerer and handed to the sorcerer the magic grass'. Similarly, 'The sorcerer revealed to the princess the old and wise king's location. The sorcerer gave to the princess a spell to sleep the lord of evil' can be rewritten as 'The sorcerer revealed to the princess the old and wise king's location and gave to the princess a spell to put the lord of evil to sleep.'

(3) Employing pronouns. If in the same sentence a reference to an actor is repeated, the second instance can be substitute with a pronoun. Applying this rule, we can transform the previous examples into the following sentences: 'The princess returned to the sorcerer and handed him the magic grass' and 'The sorcerer revealed to the princess the old and wise king's location and gave her a spell to put the lord of evil to sleep.' These changes improve the output:

> The old and wise king was at the castle. The princess was a brave person. The princess loved him. One morning, the lord of evil went to the castle and abducted the monarch. The lord of evil took the old and wise king to the cursed cave. The princess went to talk to the sorcerer to ask for help. The sorcerer agreed to provide the information in exchange for a cluster of magic grass that only grows in the sacred lake. The princess went in search of the plant but the guardian of the lake barred her from entering. The princess sang an ancestral song of supplication and the guardian of the lake allowed her to enter. The princess took a bunch of magic grass. The princess returned to the sorcerer and handed him the magic grass. The sorcerer revealed to the princess the old and wise king's location and gave her a spell to put the lord of evil to sleep. Wasting no time, the princess went to the cursed cave. The princess invoked the spell and the lord of evil fell into a deep sleep. The princess released the sovereign. The end.

In a similar way, it is possible to develop tools to change the tense of verbs, narrate the tale from different characters' perspectives (see the description of CURVESHIP in Section 13.1), and so on.

5.2 TALE-SPIN

Problem-solving has been a widely used approach to automatic narration over the past fifty years. Many systems that explore new methods for setting and solving problems have been developed. In the following we describe the central characteristics of one of the most representative of these systems.

TALE-SPIN was developed in the mid-1970s by James R. Meehan. In this story-teller the user provides the initial state of the world; that is, the user decides who are the characters participating in the tale, the kind of relationships they have, the goal they want to achieve, and so on. Then the system generates a story by describing how these characters reach this goal. This is an example of a story generated by TALE-SPIN:

Joe Bear and Jack Bear

Once upon a time, there were two bears named Jack and Joe, and a bee named Sam. Jack was very friendly with Sam but very competitive with Joe, who was a dishonest bear. One day, Jack was hungry. He knew that Sam Bee had some honey and that he might be able to persuade Sam to give him some. He walked from his cave, down the mountain trail, across the valley, over the bridge, to the oak tree where Sam Bee lived. He asked Sam for some honey. Sam gave him some. Then Joe Bear walked over to the oak tree and saw Jack Bear holding the honey. He thought that he might get the honey if Jack put it down, so he told him that he didn't think Jack could run very fast. Jack accepted this challenge and decided to run. He put down the honey and ran over the bridge and across the valley. Joe picked up the honey and went home.

(Meehan 1976, 38)

TALE-SPIN works in three modes. The first is referred to by the author as Verbose Mode. In this kind of operation, the system generates as an output a report that includes everything that happens inside the story-world. You can picture it as a kind of log file that makes it possible to study in detail the process that the program follows while developing a tale. Mode 2 is named Not-so-verbose Mode. In this case, the system produces a story that excludes from the text all irrelevant events. Mode 3 is referred to as Prepackaged Plot. Here, the storyteller employs a predefined group of initial states and constraints to generate stories with morals. For instance, if the user selects to develop a story around the message 'Never trust flatterers', the final tale must include at least two characters, one that flatters and one who is flattered, and a scene where the flattered actor loses something he appreciates as a result of the flattery.

Meehan established fourteen characters that could participate in a narrative: bear, bee, boy, girl, fox, crow, ant, hen, lion, dog, wolf, mouse, cat, goat, canary. In a similar way, the system has a list of characters' names that are assigned at random; predefined objects, such as berries, flowers, rivers, and worms; and possible locations, such as mountains, meadows, valleys, caves, and nests.

5.2.1 Initial state

When the program starts, the user selects one or more characters to participate in the tale; then the system creates an appropriate environment for each of them. For instance, if the user chooses a bee as a character, the storyteller knows that it requires a beehive as home, which is located at a tree, and that the beehive has honey that belongs to the bee; a bear needs a cave to live, which is in a mountain; and so on. At the beginning, all characters are located automatically inside their predefined homes. In this way, rather than employing a static predefined story-world, the system creates one at runtime based on the necessities of the current story; so if there is no bear in the tale, the cave and the mountains are not created. Meehan provided the story-teller program with detailed information about how to organize the story-world. For instance, it is not allowed to have a beehive inside a river. If two locations are not connected (e.g. the cave and the tree that holds the beehive), the program inserts the necessary elements into the story-world (e.g. a valley) to allow characters to move from one site to another. Nevertheless, this dynamic story-world can only include those elements previously defined by the author. Similarly, information like the kind of food that each creature eats, the sort of objects that they might own, or how they move around (e.g. walking, flying) is given to the program.

The user decides on a problem to overcome; the protagonist might face new inconveniencies or cause troubles for other characters during her attempts to sort it out. There are four main problems that characters must deal with: to be hungry, thirsty, horny, or tired.

5.2.2 Relationships and personalities

Relationships between actors are established by the user; there are seven types: competition, dominance, familiarity, affection, trust, deceit, and indebtedness. They can have a value between −10 and +10. A relationship of trust set to −10 represents a total distrust between characters while a value set to +10 represents a complete trust between characters. So if the bear requires to find some berries to eat and he trusts the bee, then he will ask the bee for information about how to obtain food. In a similar way, the user can assign personalities to the characters; there are four types: kind, vain, honest, or intelligent, which can also have values between −10 and +10. Thus, if the bee is a dishonest character, the bee may fool the bear about where to find berries if that is a way to gain a benefit. Relationships and personalities are primarily used as preconditions to perform actions and plans.

5.2.3 Rules of inference

TALE-SPIN includes a set of rules, known as inferences, which are checked each time the current story-state is modified, namely, when something happens in the story-world. There are three main types. The first determines the consequences of characters' behaviours: for example, if Anne eats cheese, then the cheese no longer exists and Anne is not hungry; if Adam strikes John, then John is hurt; if Mary pushes Carmen into the river, then Carmen is in the river; if Amanda takes her books home, then Amanda is at home, Amanda's books are also at home, and Amanda knows that her books are at home. The second type makes characters respond to contextual situations. For instance, if Wilma is in the river, then she wants to get out (either swimming, flying, or asking for help depending on her skills); if Paul is hungry and sees food, then he wants to eat it; if Paul fails to get that food he gets sick. The third type represents characters' reactions to new knowledge; most of the time those reactions trigger either relationships between characters or new goals to pursue. For instance, when Arthur finds out that Helen thinks that Arthur is brilliant, then Arthur likes and trusts Helen; when Daniel finds out that Amadeus steals Daniel's bread (or anything he owns), then Daniel dislikes Amadeus; when John finds out that Bill thinks that John cannot run very fast, then John wants to run fast (if he feels competitive towards Bill). All those inferences occur when the program is running and are helpful to generate more dynamic tales.

5.2.4 Goals

TALE-SPIN includes goals to meet physical needs, such as resting, becoming healthy, satisfying hunger; goals to move around the story-world, such as moving close to someone, getting away from someone, figuring out a route to a location; goals to persuade other characters to act in certain ways; goals to communicate information, to find out information, to acquire an object; and so on. A goal-structure consist of a set of methods, known as planboxes, whose purpose is to provide characters with various options to reach the goal. Each planbox has associated its own set of preconditions and plans. Here is an example. DPROX (X, Y, Z) is a procedure that character X uses to get Y near Z (where X, Y, and Z are variables). Y and Z can represent characters, objects, or locations. If X is the same as Y, the system interprets that X wants to be near Z. So DPROX satisfies the goal to locate Y near Z. To illustrate its usefulness, you can imagine a tale where John bear wants to put a delicious worm near Wilma bird to convince her of giving him some information. The program would call on DPROX to solve the problem. DPROX has three planboxes:

- Planbox-1: character X tries to move Y to Z;
- Planbox-2: character X gets character Y to move himself to Z;
- Planbox-3: character X gets someone else, character P, to move Y to Z.

TALE-SPIN first tries to put Planbox-1into practice; if the plan fails, then it tries to perform Planbox-2; if it fails, then Planbox-3 is executed; if all plans fail, the goal cannot be achieved and therefore the tale cannot be told. Planbox-1 is organized as follows:

Planbox-1: character X tries to move Y to Z
PRECONDITIONS
If X is different from Y then

(X moves near Y) and (X grabs Y).
X finds out where Z is.
X finds out where it itself is.
X figures out a route to Z.
PLAN
X moves Y near to Z.
If X had picked up Y then X drops Y.

The first precondition indicates that if character X wants to move an object or another character (represented by the variable Y), first X should be near Y, then X grabs Y. The second precondition requires that character X figures out where Z is. The variable Z can represent a location or another character. So the system triggers the goal of getting the necessary information to fulfil this requisite. The third precondition demands that character X knows where X itself is. This requirement attempts to address situations where a character is lost or disoriented and, therefore, needs to discover its own location. The last precondition triggers the goal of figuring out a route that allows character X to move from its actual position to the site linked to Z. Once all preconditions are fulfilled, the plan associated to Planbox-1 is executed. First, the actual action of moving takes place, namely, the location of character X and Y is updated; finally, character X lets Y go.

When the program starts, the user choses a goal to be pursued. Goals are composed of planboxes which trigger new goals that have their own planboxes and so on. Inference rules might also trigger new goals; some other goals reoccur, like being hungry, horny, thirsty, or tired; that is, the system reactivates them automatically after a certain period of story-time has passed.

5.2.5 Outputs

TALE-SPIN employs a program named MUMBLE, written by Meehan in a day, to transform the information registered in the story-structures into English. Not surprisingly, Meehan was not satisfied with the results. Therefore, sometimes, as in the example presented at the beginning of this section, Meehan 'translates' by hand some of the products created by TALE-SPIN: 'so, I present here a translation done by hand, for ease of reading. All the events in the story were produced by the program; only the English is mine' (Meehan 1976, 36). Thus, it is not clear to us which part of the final text is written by a human and which part is written by the system. The author provides an interesting example that compares a description generated by MUMBLE and the same description written by Meehan in the form of dialog.

Text generated by MUMBLE:

(a) TOM BEAR ASKS GEORGE BIRD WHETER IF TOM BEAR GIVES A WORM TO GEORGE BIRD THEN GEORGE BIRD WILL ASK IVAN BEE

WHETER IF TOM BEAR GIVES A FLOWER TO IVAN BEE THEN IVAN BEE WILL GIVE THE HONEY TO TOM BEAR.

(b) GEORGE BIRD TELLS TOM BEAR THAT GEORGE BIRD WILL ASK IVAN BEE WHETHER IVAN BEE WILL GIVE THE HONEY TO TOM BEAR IF TOM BEAR GIVES A FLOWER TO IVAN BEE.

A hand translation in the form of a dialog:

(a) George, would you ask Ivan if he'll trade me his honey for a flower? I'll give you a worm if you do.
(b) Sure, Tom.

This case clearly illustrates how difficult it is for an automatic storyteller to generate an adequate text for a narrative; it also highlights the need to consider both the process and the outputs when evaluating a system.

5.2.6 Horror stories

Meehan gives the term 'Horror stories' to stories that illustrate inaccuracies or mis-calculations in the automated narrator. In his dissertation, he includes a chapter entitled 'Mis-spun tales', where he describes some of TALE-SPIN's horror stories. The following shows a good illustrative example:

Henry Ant was thirsty. He walked over to the river bank where his good friend Bill Bird was sitting. Henry slipped and fell in the river. Gravity drowned.

In an early version of TALE-SPIN, Meehan gave gravity the properties of a character. When this story was generated, the system included the following rules:

1. If A moves B to location C, then A and B are placed in location C.
2. When a character is in a river, she wants to get out to avoid drowning. Thus, if the character has legs, then she swims out; if she has wings, then she flies away; if the character has friends, then she asks for help.

In this story, because Henry slipped and fell, gravity is what moves Henry into the river, so both ended in the water (according to rule 1 above). Then, as a result of not having neither legs, wings, nor friends, gravity drowned (according to rule 2 above).

Despite the limitations of its text generation, TALE-SPIN is a great storyteller that pioneered concepts such as the use of characters' goals to drive story generation, the development of narratives around messages like 'Never trust flatterers', and the creation of the story-world at runtime.

5.3 Final remarks about problem-solving in narrative generation

A well-designed system based on problem-solving techniques can produce appealing short stories. Characters' aims are the forces that drives a narrative. The use of rules and a random selection of actions are useful for generating more diverse outputs. Goals are essential for maintaining the coherence of the tale and are helpful for inserting narrative elements such as suspense and tragedy. The use of nested goals allows for elaborated plots to be generated. The elements that make up this type of system can be handled in a modular way. That is, goals and actions work as building blocks to the benefit of different stories. Even a driving goal, like the one that gave rise to the example in Section 5.1, might operate as a subgoal in another plot. For instance, a love story might include an abduction episode when one of the characters has a mean personality. In this way, systems based on problem-solving exploit the ability of computers to combine components in multiple ways and thus produce original results.

The main limitation of this approach is its rigidity. All the relevant elements in a narrative, such as goals and plans, need to be defined in advance. Thus, characters never will be able to solve problems that were not anticipated by the designers.

6
Narrative generation from the author's perspective
Planning

6.1 Introduction to planning in narrative generation

In Chapters 3 to 5 we studied how to generate narratives by reporting characters' efforts to solve problems. The stories that those kinds of system produce are developed around what is known as *character goals*, that is, around the necessities and desires that a character in a story might have. However, this approach sometimes produces undesired effects. For instance, if a storyteller employs as an initial state 'Bruce was tired', the system will probably trigger a character goal for resting; as a result, it could produce the following output: 'Bruce was tired. He went to sleep. The end.' This piece is unsatisfactory even though the actor has solved his problem. In this chapter, we complement and expand the notions learnt in Chapters 3–5 by studying how storytellers employ what is known as *author goals* to avoid these kinds of trouble.

An author goal is a computer representation of a goal of a human storywriter. There are several types, for example, the aim to develop a story that illustrates a given *theme*. In this work, a theme is a grouping of story elements that belong to a particular type of narrative. Thus, a theme might be inspired by a fable, such as 'the tortoise and the hare', 'the fox and the crow'; by a description of personal anxieties, such as How am I going to find the love of my life? How am I going to succeed in my job?; or even by an account of social concerns, such as the dangers that immigrants face when crossing the border. All these situations can be classified by their distinctive characteristics, some of which can be represented in computer terms. Thus, themes capture and organize, from a computational-narrative perspective, the core elements of types of narratives like fables, personal anxieties, and social concerns, to guide the generation of stories.

For instance, the description of a theme that represents the fable of the fox and the crow includes at least two characters: a flatterer and a bragger. The flatterer must desire an object that the bragger owns. The flatterer creates a situation that allows the bragger to show off; as a result, the bragger loses his object. Because the characteristics that make up a theme are general, there are several potential tales that would fulfil it. Here is another example. A theme about immigrants being trafficked requires three types of character: immigrants, human traffickers, and border officials. The narrative must include an initial situation that forces people to emigrate, a

An Introduction to Narrative Generators. Rafael Pérez y Pérez and Mike Sharples, Oxford University Press.
© Rafael Pérez y Pérez and Mike Sharples (2023). DOI: 10.1093/oso/9780198876601.003.0006

situation where emigrants contact human traffickers to cross the border and finally an outcome that involves border officials (Chapter 14 discusses the use of automatic narrative generation to express social and political concerns). Thus, a theme groups together those features that allow a narrative to be identified as belonging to a given type (or genre). It is a prerogative of the system designers to define these features and types. The construction of narrative elements, for example suspense or flashback, is another example of author goals that we will discuss in Chapter 7.

Early artificial intelligence (AI) research in problem-solving techniques involved the resolution of relatively simple challenges. When researchers increased the complexity of the tasks that AI systems had to deal with, they developed new methods that made it possible to cope with more elaborated environments. 'These methods focus on ways of decomposing the original problem into appropriate subparts and on ways of recording and handling interactions among the subparts as they are detected during the problem-solving process. The use of these methods is often called planning' (Rich & Knight 1991, 330). As we will see in this chapter, the management of author goals and character goals requires the use of planning methods.

6.2 Developing stories around themes

We define a *story* that illustrates a theme as a sequence of events where characters face one or more problems that prevent them from reaching their goals. It is organized in four sections: introduction, development, resolution, and closure. The introduction familiarizes the reader with the main characters and the settings of the tale, the development introduces one or more scenes where the lives of the main characters intertwine usually in troublesome circumstances, the resolution depicts the climax of the story and explains how all the problems are sorted out, and the closing offers a conclusion for the narrative. Figure 6.1 shows a representation of a theme structure based on this definition.

Each attribute of a theme's structure is filled in with one or more *descriptions of situations* that are required to occur in the tale. For instance, the theme 'Do not give up because you never know when things will turn around' looks as follows (Character-A and Character-B are variables that represent any valid character; Location is a variable that represents any valid location):

Characters: Character-A is a female, Character-B is a male.

Introduction: Character-A and Character-B live in the same Location. Character-B fancies Character-A.

Development: The story progress in a way that Character-A ends getting upset with Character-B.

Resolution: A critical event occurs that causes Character-A to feel grateful towards Character-B.

Closing: The story ends with Character-A and Character-B having a romantic event.

These general descriptions offer guidelines about how to progress a narrative. At the same time, they provide detailed information for the generation of the piece, such

Theme Structure
Characters:
Introduction: Development: Resolution: Closing:

Figure 6.1 Representation of a theme structure

as the parts that make up a story, the core features of the main passages, the order of events, the main characters, how conflicts between characters evolve. For instance, in this case, conflicts progress as follows: first, because Character-B fancies Character-A, a potential problematic situation associated with 'be rejected' arises; next, Character-A gets upset with Character-B, generating a conflict; finally, Character-A is grateful towards Character-B, resulting in a change of Character-A's attitude and therefore solving the problem. *In this way, a theme introduces a set of constraints that guarantees the development of coherent and interesting plots around specific topics.*

6.2.1 Employing themes to generate a narrative

Computer storytellers include a collection of actions, goals, objects, locations, and characters that are represented in the system. This information is stored as data structures in a repository known as its *knowledge base* (in contrast with human memory, these repositories allow information to be preserved permanently). Let us use a metaphor to illustrate this idea. The knowledge base can be pictured as a set of shelves, each of which holds boxes. As shown in Figure 6.2, inside these boxes there are several paper cards, each one representing a story-action structure, a goal-structure, a character-structure, or the name of a character, object, or location. The theme is represented as a piece of paper with written guidelines that indicate the characteristics of the cards necessary to develop a story. Some of the boxes represent data structures and others represent elements to be used in those structures. The box labelled 'Characters' includes the names of characters in the story and it is used to assign a value to any variable in a data structure that represents then name of a character. The box labelled 'Objects' includes the names of multiple objects and it is used to assign a value to any variable in a data structure that represents an object. The box labelled 'Locations' includes the names of diverse locations and it is used to assign a value to any variable in a data structure that represents a location. During the generation of a narrative, the system can only employ cards that are part of the knowledge base. If necessary, new items can be added to the boxes. The designer of the system decides the characteristics of the theme, as well as the number of boxes and cards that the knowledge base contains.

In this way, a storyteller system searches in the knowledge base for structures that are useful for building concrete scenes that illustrate the theme. This process is called *instantiation* and works as follows. The description in each attribute of the theme

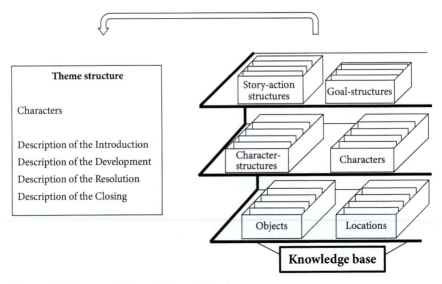

Figure 6.2 Representation of a knowledge base

structure—Introduction, Development, Resolution, and Closing—functions as an indicator to match one or more structures in the knowledge base; this indicator is known as an *index*. In other words, the index specifies the features of the structures that the program is looking for in the knowledge base. For instance, in this exercise the Introduction specifies that the storyteller must retrieve an action or a goal that, as a result of executing it, places Character-A and Character-B living in the same Location. A simple story action describing that Character-A and Character-B decide to move to the same Location satisfies the requirement. Besides this, it is necessary to find an action or goal that, as a result of executing it, one of the character fancies the other. Again, as shown in Figure 6.3, the system needs to look in the knowledge base for a structure that satisfies these requirements. For instance, the action 'First-character finds attractive Second-character' results in one actor fancying the other.

Note that the variables representing actors in story-action structures are referred to as First-character, Second-character, and so on, while variables representing actors in the theme structures are referred to as Character-A, Character-B, and so on. Although the theme and the story action both employ characters, each of these structures represent different aspects of a narrative. Therefore, for the sake of clarity, their variables representing actors have different names.

Next, the attribute Development needs an action or a goal that results in one of the characters becoming upset with the other. The same search and retrieval process is performed again. This method is repeated for all attributes. If the knowledge base includes a rich set of elements, such as actions, goals, and locations, the same theme can be instantiated in multiple ways producing very different kind of tales. Similarly, if the storyteller works with two or more themes, the number of possible narratives that it can develop increases. Thus, *developing a story consists of instantiating the structures that compose a theme.*

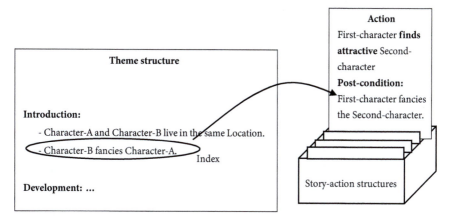

Figure 6.3 Illustration of how a theme structure's element is used as an index to match a story action

6.2.2 Representing themes as data structures

We now explain how the instantiation process works when a theme is represented in terms of data structures. Figure 6.4 shows 'Do not give up because you never know when things will turn around' in this schematized format. We employ doble-line boxes to represent data structures. The figure has a particular feature: except for the theme structure, the rest of structures, which represent characters, goals, and actions, are only partially defined; that is, only some of their attributes are shown. Their purpose is to serve as indexes (or indicators) for searching similar structures in the knowledge base to instantiate the theme. In other words, they work as search terms to be matched against items in the knowledge base. Thus, when the attribute Introduction is processed, the system searches for a story-action structure that includes as a post-condition the fact *Character-A and Character-B live in the same Location.*

Once it finds a story-action structure that satisfies this requirement of two individuals residing in the same place, the program retrieves this structure and uses it to instantiate the theme and, in this way, to step forward the tale. If the program matches two or more options, one of them is selected at random. Similarly, when processing Development, the system looks for a goal-structure whose plan includes the action *Character-A gets upset with Character-B* to progress the plot, and so on. Arrows in Figure 6.4 indicate how partially defined structures are linked to attributes of the theme structure. Arrows also show the order in which each action or goal must be presented in the scene; for example, in the Introduction, first the characters are placed in Location, then one of them fancies the other.

You might wonder why in Figure 6.4 the Introduction employs actions' post-conditions as indexes while the Development, Resolution, and Closing use goal-structures' plans as indexes. This is a design decision that depends on the features of the theme. In this case, we want to demonstrate how both types of structures can be useful in progressing a tale. Other features, such as preconditions, name of goals,

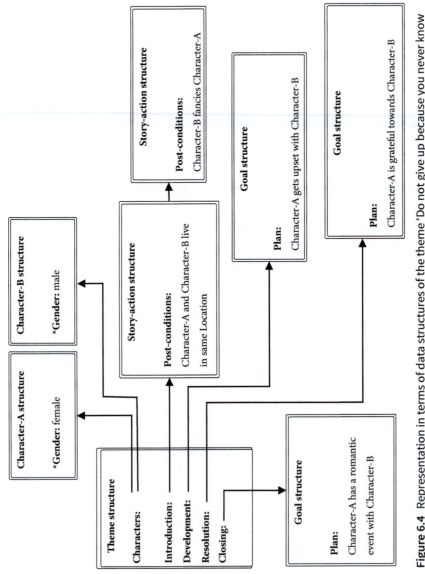

Figure 6.4 Representation in terms of data structures of the theme 'Do not give up because you never know when things will turn around'

or a combination of them, can also be used as indexes. In the subsequent sections we present the general design of a storyteller system that employs themes to develop narratives.

6.3 Designing a storyteller based on themes

The following human-written text illustrates the theme 'Do not give up because you never know when things will turn around':

The plot by the river

Carmen and Juan were neighbours in a small town next to the mountains. Juan thought that Carmen was beautiful. Carmen wanted to buy the plot by the river. Without knowing Carmen's intentions, Juan decided to buy the same piece of land. The owner, Old Tom, decided to sell it to Juan. Carmen was upset with Juan about it. A few days later Carmen went for a walk in the mountains when she tripped and hit her head. While riding his bike Juan found her and carefully healed each of her wounds. Carmen was very grateful to Juan and she decided to forget the unpleasant incident about the plot. That night they dined by candlelight.

We employ this narrative to guide the design of the storyteller; that is, our system must be able to reproduce this tale, or a similar one. The first step is to build the knowledge base; thus, it is necessary to identify the core features of the story world. In this case, *The plot by the river* includes three characters: Carmen, Juan, and Old Tom; two locations: a small town next to the mountains and the mountains; a story object: the plot by the river (for facility, the plot is considered an object). Besides this, characters in this tale get injured, have emotional relations with each other, desire objects, and own objects.

Figure 6.5 shows a character-structure that represents all these features. It includes six attributes: name, gender, location, health condition, emotional relations (between characters), desired objects, and objects owned. The value of all these attributes is assigned at runtime.

```
Character structure

*Name:
*Gender:
*Location:
*Health-Condition:
*Emotional-relations:
*Desired-objects:
*Objects-owned:
```

Figure 6.5 The character structure for the tale *The plot by the river*

Table 6.1 Characters' attributes for the story *The plot by the river* and their possible values

Attribute	Possible values
*Name	Carmen, Juan, Old Tom, Andrea, Uncle Albert, Laura
*Gender	Male, female, non-binary
*Location	Small town by the mountains, fishing village, mountains, forest trails, path by the river
*Health-condition	Healthy, injured, ill
*Emotional-relations	None, friends with, grateful to, fancies, flirts with, in love with, upset with, hates
*Desired-Objects	None, tractor, plot by the river, car, horse
*Objects-owned	None, tractor, plot by the river, car, horse

Table 6.1 shows the possible values that can be assigned to each of the characters' attributes. The table does not only include those elements used in *The plot by the river* but also other options that can be exploited in future pieces. In this way, the next tale produced by our system might describe Andrea falling in love with Laura, or Uncle Albert owning a horse, among other possibilities.

The analysis of structures' attributes and their possible values are essential for the design of a storyteller because they determine the content of the knowledge base. In this way, continuing with the metaphor introduced earlier, the column 'Attribute' in Table 6.1 indicates that the system requires six boxes labelled as follows: Name (of characters), Gender, Location, Health-condition, Emotional-relations, and Objects (we use one box for Desired-objects and Objects-owned because both attributes employ the same possible values). The column 'Possible values' specifies the content of each box: for example, the box Name includes six cards, each with one name: Carmen, Juan, Old Tom, Andrea, Uncle Albert, and Laura; the box Gender includes three cards, one for male, one for female, and one for non-binary.

Table 6.2 shows the thirteen actions that characters perform in *The plot by the river*. Their function is to modify the story world, in this case to modify the characters' and objects' attributes. All these actions involve two participants, First-character and Second-character, which represent variables that can be substituted with any character in the tale. Some actions include objects or locations, represented by the variables Object-for-this-action and Location-for-this-action, which can be substituted by any valid value shown in Table 6.1. For the sake of simplicity, in this example we do not define preconditions (which were studied in Chapters 3 to 5). However, a complete design must contemplate those elements. As in the previous cases, it is possible to add as many deeds as necessary. Continuing with the metaphor, Table 6.2 shows that the knowledge base requires a box labelled story-action structures that contains thirteen cards describing each of the deeds. Note that in Figure 6.3 the post-condition of the action 'First-character **finds attractive** Second-character' is portrayed as a general description ('First-character fancies the Second-character') while in Table 6.2, the same action's post-condition is shown in algorithmic terms

Table 6.2 Alphabetically ordered list of story-actions employed in *The plot by the river*

	Story-action structures	
	Action	Post-conditions
1	First-character and Second-character **are neighbors in** Location-for-this-action	First-character's attribute *Location is set to the current value of the variable Location-for-this-action. Second-character's attribute *Location is set to the current value of the variable Location-for-this-action.
2	First-character **cures** Second-character	Second-character's attribute *Health-condition is set to 'healthy'.
3	First-character **finds by accident** Second-character	First-character's attribute *Location is set to the same value of Second-character's attribute *Location.
4	First-character **finds attractive** Second-character	First-character's attribute *Emotional-relations is set to 'fancies Second-character'.
5	First-character **gets upset with** Second-character	First-character's attribute *Emotional-relations is set to 'upset with Second-character'.
6	First-character **goes to** Location-for-this-action	First-character's attribute *Location is set to the current value of Location-for-this-action.
7	First-character **goes to look for** Second-character	First-character's attribute *Location is set to the same value of Second-character's attribute *Location.
8	First-character **has an accident**	First-character's attribute *Health-condition is set to 'injured'.
9	First-character **has a romantic event with** Second-character	First-character's attribute *Emotional-relations is set to 'flirt with Second-character'. Second-character's attribute *Emotional-relations is set to 'flirt with First-character'.
10	First-character **is grateful towards** Second-character	First-character's attribute *Emotional-relations is set to 'grateful towards Second-character'.
11	First-character **owns** Object-for-this-action	The Object-for-this-action is included in First-character's attribute *Objects-owned.
12	First-character **sells** Object-for-this-action to Second-character	The Object-for-this-action is eliminated from First-character's attribute *Objects-owned and it is included in Second-character's attribute *Objects-owned.
13	First-character **wants to buy** Object-for-this-action	The Object-for-this-action is included in First-character's attribute *Desired-objects.

('First-character's attribute *Emotional-relations is set to 'fancies Second-character'). Transforming general descriptions into algorithmic definitions is one of the main challenges of designing computer models of writing.

Finally, we need to include in our knowledge base the story goals; as we learned in Chapter 3, they are structures that employ actions and other goals to accomplish specific ends. Figure 6.6 shows the three goals that are used in this example.

The goal *Object envy* represents a situation where two characters want to buy the same object without being aware of the intention of the other. The goal involves Character-A, Character-B, Character-C, and Object-goal-1; the value of all these variables is assigned at runtime. The first precondition states that Character-C must own Object-goal-1. The second and third preconditions indicate that both Character-A and Character-B want to buy Object-goal-1. The plan indicates that once the preconditions are satisfied, the action *Character-C sells Object-goal-1 to Character-B* is executed and, as a result, *Character-A gets upset with Character-B* is performed.

The goal *Helping an injured person* represents a situation where someone helps an injured character. The goal requires two participants, Character-A and Character-B, and a location represented by Location-goal-2; the value of these three variables is established at runtime. The preconditions require that Character-A decides to go to Location-goal-2 and there she suffers an accident. Then the plan states that

---

Goal-structure 1

Name of the goal: Object envy
***Characters:** Character-A, Character-B, and Character-C.
***Object:** Object-goal-1
Preconditions:
- Character-C's attribute *Objects-owned must include Object-goal-1.
- Character-A's attribute *Desired-Objects must include Object-goal-1.
- Character-B's attribute *Desired-Objects must include Object-goal-1.
Plan:
- Character-C sells Object-goal-1 to Character-B.
- Character-A gets upset with Character-B.

Goal-structure 2

Name of the goal: Helping an injured person.
***Characters:** Character-A, Character-B.
***Location:** Location-goal-2
Preconditions:
- Character-A goes to Location-goal-2.
- Character-A has an accident.

Plan:
- Character-B finds by accident Character-A.
- Character-B cures Character-A.
- Character-A is grateful towards Character-B.

Goal-structure 3

Name of the goal: Going on a date.
***Characters:** Character-A, Character-B
Preconditions:
- Character-A and Character B have positive feelings towards each other, i.e. the emotional relations from Character-A to Character-B, and vice versa, are set to 'friends with', 'grateful to', 'fancies', 'flirts with', or 'in love with'.
Plan:
- Character-A has a romantic event with Character-B.

Figure 6.6 Goal-structures for the story *The plot by the river*

Character-B accidently finds Character-A and then he cures her wounds. As a result, Character-A is grateful towards Character-B.

The goal *Going on a date* describes a situation where two characters go for a date. The goal employs two participants, Character-A and Character-B, who are instantiated at runtime. The preconditions require that Character-A and Character-B have positive feelings towards each other, that is, that the emotional relations from Character-A to Character-B, and vice versa, are set to 'friends with', 'grateful to', 'fancies', 'flirts with', or 'in love with'. The plan describes that Character-A goes for a romantic event with Character-B.

The number of goals that can be added is endless. Continuing with the metaphor, this analysis shows that the system requires a box labelled goal-structures that contains three elements. Figure 6.7 shows the whole knowledge base for this example in terms of boxes and cards. It includes three boxes representing data structures and six boxes representing elements used in the tale.

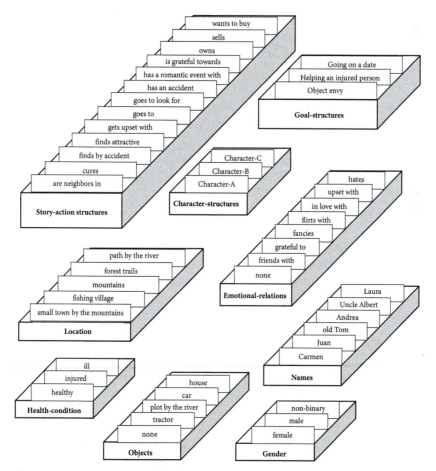

Figure 6.7 Representation of the knowledge base for *The plot by the river*

In summary, a theme structure is a collection of partially defined characters, goals, and action data structures. The knowledge base includes:

- Character-structures (representation of characters as data structure);
- Story-actions structures (representation of story actions as data structures);
- Goal-structures (representation of characters' goal as data structure); and
- List of names of characters, objects, health condition, gender, emotional relations, and locations.

A more elaborated system might include as part of its database a set of themes to be developed.

6.4 Generating a story based on a theme

Next, we develop a computer program that narrates a story based on our theme. It is beyond the scope of this book to explain how this program can be implemented in a specific computer language; instead, we describe how it works in general. The generation process includes seven main steps:

1. Choosing a theme structure.
2. Creating the characters' structures of the theme.
3. Instantiating the Introduction.
4. Instantiating the Development.
5. Instantiating the Resolution.
6. Instantiating the Closing.
7. Transforming the data structures into English.

1. Choosing a theme structure. A storyteller might include two or more theme structures in its knowledge base. Thus, the first step is to choose which of them will be developed. This decision can be taken either by the user of the system or by the program itself, for example selecting one at random. For this exercise, the program picks the only available theme.

2. Defining characters. This process creates the structures for Character-A and Character-B. The theme requires that the attribute *Gender of Character-A be set to 'female' and the attribute *Gender of Character-B be set to 'male'. This is an arbitrary decision; it is possible to build a theme with two males, two females, two non-binaries, or any combination of them. Because the values of the rest of the attributes for the two characters are not specified, the program must choose ones for each of them. The value of the attribute *Name is chosen at random (although this procedure might be designed to constrain the selection of names based on the value of *Gender). Thus, for this exercise *Name of Character-A is set to 'Carmen' and *Name of Character-B is set to 'Juan'. For simplicity, the rest of the attributes are assigned, by default, the first element of the column Possible Values in the Table 6.1. In this way, *Location is set to 'small town by the mountains', *Health-condition is set to 'healthy', *Emotional-relations is set to 'none', *Desired-objects is set to 'none', and *Objects-owned is set to 'none'.

3. Instantiating the Introduction. As shown in Figure 6.4, the Introduction includes two story-action structures and therefore the process is divided in two parts.

Step 1: The index of the first structure is the post-condition *Character-A and Character-B live in Location*. Let us break it down. The system needs to find an action that involves two characters and one location. The description 'lives in' indicates that at least one of the action's post-conditions must produce that both characters are placed in Location; that is, Character-A's attribute *Location must be set to Location and Character-B's attribute *Location must also be set to the same Location. These are the constraints that should be satisfied. The program launches a search in the knowledge base to trace all those story-action structures whose post-conditions match the index; it retrieves at random *First-character and Second-character are neighbors in Location-for-this-action* shown in Table 6.2-1.

As we mentioned earlier, variables representing actors in story-action structures are referred to as First-character, Second-character, and so on, while variables representing actors in the theme structures are referred to as Character-A, Character-B, and so on; similarly, variables representing locations in the story-action structures are referred to as Location-for-this-action while variables representing locations in the theme structures are referred to as Location. Because theme structures and story-action structures use variables to represent characters, locations, and objects, the storyteller must work out the correct relation between the theme structure's characters and the story-action structure's characters, the theme structure's locations and the story-action structure's locations, and the like. In this case, as shown in Figure 6.8, the program figures out that the variable Character-A corresponds to the variable First-character in the action, the variable Character-B corresponds to the variable Second-character in the action, and the variable Location corresponds to the variable Location-for-this-action.

The system substitutes the value Carmen for Character-A and Juan for Character-B. Because Location has not been assigned a value yet, the program selects at random 'small town by the mountains'. In this way, the action *Carmen and Juan are neighbours in the small town by the mountains* is performed, and Carmen and Juan's attributes *Location are set to 'small town by the mountains'.

Step 2: The program probes the knowledge base to seek all those actions whose post-conditions match the second index *Character-B fancies Character-A*; that is, the action should include as post-condition that Character-B's attribute *Emotional-relations is set to 'fancies Character-A'; it retrieves *First-character finds attractive Second-character* shown in Table 6.2-4. Again, the system figures out that Character-A links to First-character and Character-B links to Second-character, and then Character-A and Character-B are substituted with Carmen and Juan, respectively.

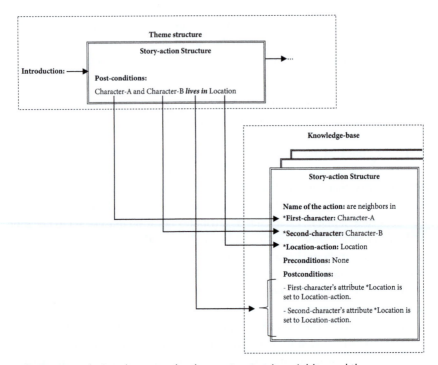

Figure 6.8 Relations between the theme structure's variables and the story-action structure's variables

The deed *Juan finds attractive Carmen* is performed and as a result Juan's attribute *Emotional-relations is set to 'fancies Carmen'. In this way, the Introduction is instantiated.

4. Instantiating the Development. As shown in Figure 6.4, the Development includes a goal-structure that employs the plan's action *Character-A gets upset with Character-B* as index. The program searches the knowledge base and this time it matches the goal *Object envy*. As shown in Figure 6.6, this goal includes, besides Carmen and Juan, a third character and an object referred to as Character-C and Object-goal-1, respectively. The system creates a new character-structure to represent Character-C. It chooses at random 'Old Tom' for *Name, and the rest of the attributes are set to the first value of the column Possible Values in Table 6.1. Then 'The plot by the river' is chosen at random as Object-goal-1. Because we have defined the value of all the variables involved, the goal can be depicted as follows (cf. Figure 6.6):

GOAL-STRUCTURE 1

Name of the goal: Object envy
Characters: Carmen, Juan, and Old Tom
Object: the plot by the river
Preconditions
   o Old Tom's attribute *Objects-owned must include 'the plot by the river'.
   o Carmen's attribute *Desired-objects must include 'the plot by the river'.
   o Juan's attribute *Desired-objects must include 'the plot by the river'.

Plan

- o Old Tom sells the plot by the river to Juan.
- o Carmen gets upset with Juan.

As we saw in Chapter 3, the program needs to satisfy the goal's preconditions and then execute the plan. Thus, the system runs the procedure to satisfy the preconditions. The first indicates that Old Tom must own the plot by the river. Because Tom's attribute *Objects-owned has been set to 'none' the program looks for an action or goal that helps to satisfy the precondition. It finds that the consequences of action *Old Tom **owns** the plot by the river* shown in Table 6.2-11 solves the problem. The deed is performed and as a result 'the plot by the river' is included in Old Tom's attribute *Objects-owned. Something similar happens for the second and third preconditions. Carmen's attribute *Desired-objects has been set to 'none'; the system finds that as result of executing the action *Carmen **wants to buy** the plot by the river* shown in Table 6.2-13, the precondition is fulfilled; that is, 'the plot by the river' is added to Carmen's attribute *Desired-objects. Finally, Juan performs the same action and 'the plot by the river' is added to his attribute *Desired-objects.

Because all preconditions are fulfilled, the goal's plan is executed. The system performs the action *Old Tom sells the plot by the river to Juan* shown in Table 6.2-12. As a result, Juan owns the plot; that is, the piece of land is eliminated from Old Tom's attribute *Objects-owned and from Juan's attributes Desired-objects; then, it is added to Juan's attribute Objects-owned. Lastly, the action *Carmen gets upset with Juan* shown in Table 6.2-5 is performed. As a result, Carmen's attribute *Emotional-relations is updated with 'upset with Juan'.

5. Instantiating the Resolution. The Resolution includes a goal-structure that employs the plan's action *Character-A is grateful towards Character-B* as index. The program searches the knowledge base and it matches the goal-structure *Helping an injured person*. The goal-structure includes two characters and one location: Character-A, Character-B, and Location-goal-2. Character-A is set to Carmen, Character-B is set to Juan, and the system selects at random 'mountains' as the value for Location-goal-2. At this point, the goal-structure looks as follows:

GOAL-STRUCTURE 2

Name of the goal: Helping an injured person
*Characters: Carmen, Juan
*Location: Mountains
Preconditions

- o Carmen goes to the mountains.
- o Carmen has an accident.

Plan

- o Juan finds by accident Carmen.
- o Juan cures Carmen.
- o Carmen is grateful towards Juan.

The preconditions indicates that the actions *Carmen goes to the mountains* and *Carmen has an accident*, shown in Table 6.2-6 and 6.2-8, need to be performed. As a result, Carmen's *Location is set to 'mountains' and her *Health-condition to 'injured'. Next, the plan is executed: Juan accidentally finds Carmen, he cures her wounds, and she is grateful towards him (see Table 6.2-3, 6.2-2, and 6.2-10). As a result of performing these actions, Juan's *Location is set to the same value as Carmen's *Location, Carmen's *Health-condition is set to 'healthy', and Carmen's *Emotional-relations is set to 'gratefulness towards Juan'.

6. Instantiating the Closing. The Closing is linked to a goal-structure that employs the plan's action *Character-A has a romantic event with Character-B* as index. The program probes the knowledge base and retrieves the goal *Going on a date*. The actors Character-A and Character-B are replaced by Carmen and Juan. The goal looks as follows:

GOAL-STRUCTURE 3

Name of the goal: Going on a date
*Characters: Carmen and Juan
Preconditions
  o Carmen's *Emotional-relations must include either 'friends with', 'grateful to', 'fancies', 'flirts with', or 'in love with' Juan, and vice versa.
Plan
  o Carmen has a romantic event with Juan.

The precondition requires that both characters have positive feelings towards each other. In other words, Carmen's attribute *Emotional-relations must include either 'friends with', 'grateful to', 'fancies', 'flirts with', or 'in love with' Juan, and vice versa. The plan states that the action shown in Table 6.2-9 *Carmen has a romantic event with Juan* is performed. As a result, Carmen and Juan's attributes *Emotional-relations are set to 'flirt with Juan' and 'flirt with Carmen', respectively.

Figure 6.9 shows the story *The plot by the river* as an instantiated theme structure. Although a story can be represented as an instantiated theme or as text, the computer can only manipulate data structures.

7. Transforming the data structures into English. The last part of the generation process transforms the data structures that represent the tale into written English. As in Chapter 3, in this exercise all story actions are associated with one or more texts in the form of templates. For simplicity, all descriptions are written in past tense. For instance, the deed 'cures' shown in Table 6.2-2 might include the phrase 'First-character carefully healed each of Second-character's wounds'; the final text would look like this 'Juan carefully healed each of Carmen's wounds'. Each action can include as many templates as wished and the program will choose one at random. The designers of the system can develop routines to improve the final output; for example, they can build functions that automatically use pronouns, avoid repeating similar words, and connect two or more consecutive actions. In this way, sequences

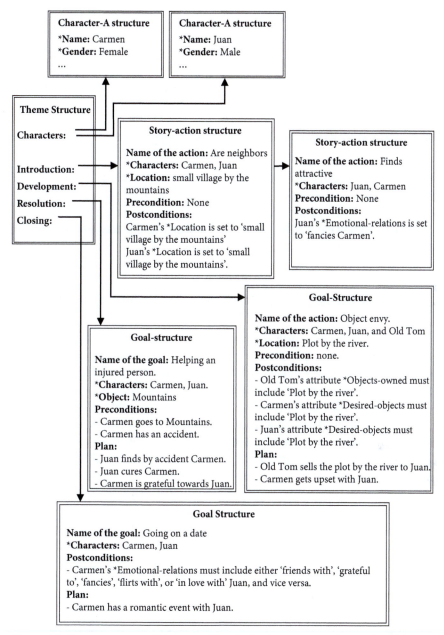

Figure 6.9 Representation of the story *The plot by the river* as an instantiated theme structure

like 'While riding his bike Juan found Carmen. Juan carefully healed each of Carmen's wounds' are transformed into 'While riding his bike Juan found her and carefully healed each of her wounds.' Additionally, we can take advantage of our knowledge about story goals and the organization of the theme structure to improve the results.

For instance, because we know that the aim of *Object envy* is that Juan buys the land before Carmen, we can include in the story expressions like 'Without knowing Carmen's intentions, Juan decided to buy the same piece of land.' Similarly, because of the Closing we know how the story ends, so it makes sense that the last section of the tale starts with expressions like 'Later that day', 'A few days later', 'The next week'.

6.4.1 A final comment about this exercise

When the program starts, many variables that represent locations, objects, and the like do not have values assigned. Through this exercise, as part of a strategy to try to generate more diverse narratives, those values are chosen at random. However, it is possible to employ more elaborate tactics. For instance, one can develop routines that choose values based on statistical information.

6.5 Final remarks about the introduction to planning

This chapter describes how goals of human writers are represented in an automatic narrative generator. We explained that an author's goal is a representation in computer terms of a goal by a human storywriter, and that a knowledge base, which characterizes the author's knowledge, is a collection of data structures that represent actions, goals, objects, locations, characters, and so on. We provide details about how stories can be developed around a structure that represents a theme, and discussed how the knowledge base is used to instantiate that theme. Finally, we considered the main steps to generate a narrative based on planning: choosing a theme structure, creating the characters' structures of the theme, instantiating in turn the Introduction, the Development, the Resolution, and the Closing, and transforming the data structures into English.

In Chapter 7 we continue our analysis on planning; we describe how to use old episodes to create new narratives and how to implement author goals like flashback and suspense, and we analyse a well-known story generator based on planning.

7
Representing the author's experience in planning

7.1 Employing known episodes to generate new narratives

All the narratives that a storyteller generates can be added to its knowledge base as a collection of episodes. Inspired by the way human memory works, we refer to this collection as episodic memory. A storyteller might employ its episodic memory to produce new tales. Let us elaborate this idea. In the example in Chapter 6, the program must figure out how to create an incident where one of the characters gets upset with the other. If this scene is saved under the category 'one character gets upset with another', the next time that the program needs to develop a similar episode, rather than spending computer resources in solving the same problem, it can simply retrieve this passage, adapt it to the new circumstances, and then use it in the tale in progress. This type of problem-solving methodology is called case-based reasoning (CBR), which is an important research area within artificial intelligence.

To illustrate how CBR in storytelling works let us imagine a theme structure that involves two characters that fall in love; next, the father of one of them gets displeased with the father of the other, generating a problem between the lovers. Thus, just as in *The plot by the river*, studied in Chapter 6, the system needs to work out what caused the upset, in this case between the two fathers. This time, the program uses the 'gets upset' description as index to retrieve from episodic memory the incident where Juan buys the plot to Old Tom and Carmen gets upset. Then the system substitutes the two fathers for Juan and Carmen; it might also change 'the plot by the river' and 'Old Tom' to some other elements that better suit the current situation; then it adds the new description to the tale in progress. In this way, the system employs its 'experience' to work through a novel problem.

The system can take advantage of the same technique to exploit events created by users. In this case, it is necessary to have a mechanism that allows users to add and modify elements in the episodic memory. Figure 7.1 represents two incidents written by these authors: a car accident and a gambling event. (For clarity, in this example we show episodes as texts; however, as we explained earlier, computers work with data structures like those shown in Figure 6.9 in Chapter 6. Thus, it is necessary to develop a program that creates such a data structure, e.g. by filling in a form to describe the happening.) In both cases, two characters take part in the incident and one of them gets upset with the other; so the storyteller can retrieve and use either of them when a

An Introduction to Narrative Generators. Rafael Pérez y Pérez and Mike Sharples, Oxford University Press.
© Rafael Pérez y Pérez and Mike Sharples (2023). DOI: 10.1093/oso/9780198876601.003.0007

EPISODIC MEMORY

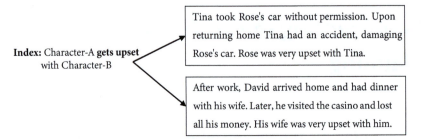

Index: Character-A **gets upset** with Character-B

Tina took Rose's car without permission. Upon returning home Tina had an accident, damaging Rose's car. Rose was very upset with Tina.

After work, David arrived home and had dinner with his wife. Later, he visited the casino and lost all his money. His wife was very upset with him.

Figure 7.1 Two incidents stored in episodic memory

story in progress requires a situation where 'Character-A gets upset with Character-B'. In this way, the scene where Old Tom sells the plot to Juan in *The plot by the river* can be switched for the car accident passage:

> *Carmen and Juan were neighbours in a small town next to the mountains. Juan thought that Carmen was beautiful. Juan took Carmen's car without permission. Upon returning home Juan had an accident, damaging Carmen's car. She was very upset with Juan. A few days later Carmen went for a walk in the mountains . . .*

However, if the system attempts to employ the description where David loses his money in the casino, things become more complicated. In this case, the husband–wife relationship helps to explain why one of the characters gets upset with the other. However, in the previous example, the fact that Carmen and Juan barely know each other is crucial for the development of plot. So, the tie between actors is an issue to be considered. But things can be worse. Let us imagine that we write an episode where one astronaut gets upset with a colleague during a mission in space; the chances that a computer system can adapt this or similar episodes to the circumstances of *The plot by the river* tale are slim. In this case, the coherence of the story in progress is jeopardized. Although diversity in the content of knowledge base promotes the production of original narratives, they also demand the development of more robust programs that can manage the associated complexity.

In summary, a storyteller based on CBR needs to include methods for storing and retrieving useful episodes, routines for adapting those episodes to novel situations, and mechanisms that make sure that the coherence of the tales is preserved.

One can combine traditional planning techniques with CBR to build more powerful storytellers. This type of hybrid systems would work as follows:

1. Use episodic memory to derive a theme structure.
2. If none of the episodes in memory are useful, or all available episodes have been utilized more than three times in the past (or any other number that the designer believes is adequate), then build a new scene from scratch using traditional planning.
3. Store the final output in episodic memory.

7.2 Creating a flashback and suspense

During plot generation, circumstances that might have an impact on the development of a narrative unexpectedly appear; for example, progressing a story in one direction rather than another might create the right conditions for inserting a flashback, or introducing a new character. Because these situations cannot be anticipated, storytellers employ what we refer to as *opportunistic rules* to detect such occasions. These sets of rules are triggered whenever their conditions are fulfilled; that is, the system constantly monitors the story in progress to try to initiate them. One common use of opportunistic rules is the production of literary devices like flashback or suspense. First, it is necessary to determine when the conditions are right for employing these devices, then describe how to implement them. We present two examples that illustrate these points.

Opportunistic-rule-Create-flashback: If the story in progress involves two lovers having an argument, then the system creates a flashback scene where one of the characters remembers a similar situation where her relatives or friends were implicated, such as a fight between her parents, her grandparents, her favourite uncle and aunt, or the parents of her best friend. This type of memory can be useful for showing doubts from one of the lovers about their relationship and in this way for complicating the plot. The storyteller can use CBR or planning techniques to build the flashback.

Opportunistic-rule-Create-suspense: If the story in progress involves an injured or ill actor, then the system delays the resolution of the current scene in order to produce suspense. The simplest way to postpone the outcome is by inserting ad hoc predefined texts. For instance, for situations where the injured character is alone, the system can add into the story, just after the injury occurs, 'The cold wind blew across Character-A's skin, foreshadowing a sad ending. It looked as if the surroundings started to disappear' (Character-A needs to be substituted with the right actor). Thus, after inserting suspense in *The plot by the river*, we end with the following text: *. . . A few days later Carmen went for a walk in the mountains when she tripped and hit her head. The cold wind blew across Carmen's skin, foreshadowing a sad ending. It looked as if the surroundings started to disappear. While riding his bike Juan found her and carefully healed each of her wounds . . .*'. To avoid repetition it is necessary to write a good number of texts. A more complex solution is to employ planning or CBR techniques to develop a delaying scene.

Opportunistic rules are also useful for altering characters' behaviour. During plot generation, some storytellers, like the one studied in Chapter 5, employ characters that get hungry or tired after a certain period of storytime. An opportunistic rule can detect this condition and trigger a goal that drives those characters to eat or rest.

The possibilities are endless. However, the use of opportunistic rules increases the complexity of the system; for example, a program might have several active author goals and character goals at the same time, and therefore it constantly needs to decide which one has priority. It is necessary to develop specialized routines that makes it possible to handle these kinds of situation.

7.3 MINSTREL

MINSTREL (Turner 1993; 1994) is a computer program that writes short stories about King Arthur and his Knights of the Round Table. The system is a case-based problem solver where past cases are stored in an episodic memory. This project represents a computer model of creative writing. Many of the explanations in this and the previous chapters are inspired by this work. This is an example of a story created by MINSTREL:

> In the spring of 1089, a knight named Lancelot returned to Camelot from elsewhere. Lancelot was hot tempered. Once, when Lancelot lost a joust, he blamed his lance and wanted to destroy it. He struck his lance and broke it.
>
> One day, a lady of the court named Andrea wanted to pick some berries. Andrea went to the woods. Andrea picked some berries. Lancelot's horse unexpectedly carried him into the woods. Lancelot was near Andrea and fell in love with her. Lancelot wanted Andrea to love him.
>
> Some time later, Lancelot's horse again carried him into the woods. There he saw Andrea kissing a knight named Frederik. Because of this, Lancelot believed that Andrea loved Frederik. Because Lancelot loved Andrea, he wanted her to love him. Lancelot hated Frederik. Lancelot's hot temper made him mad. He wanted to kill Frederik. Lancelot moved to Frederik and fought him. Frederik was killed.
>
> Andrea ran to Frederik. Andrea told Lancelot that Frederik was her brother. Lancelot wanted to take back that he wanted to kill Frederik, but he could not. Lancelot hated himself.
>
> Lancelot became a hermit to hide his shame. Frederik was buried in the woods where he died, and Andrea became a nun.
>
> (cited in Pérez y Pérez and Sharples 2004, originally published in Turner 1993, 622).

7.3.1 Author's goals and data structures in MINSTREL

Rather than employing data structures, MINSTREL represents author's goals as a set of routines programmed by the designer of the system. Examples are the goal of developing a theme, or the goal of including suspense, revenge, or deception in a story.

The elements that compose a story are represented as data structures that the author refers to as schemas. Some examples are character goals (e.g. satisfying one's hunger, finding a romantic love), representations of humans, physical objects, and emotions. Schemas can be linked to establish relationships between two or more characters, between a character and an object, between a character and some emotions. In this way, it is possible to construct elaborate scenes.

MINSTREL develops stories around six predefined themes (represented as data structures) that have the following organization:

- Introduction
- Development of the theme
    i. World facts (preconditions)
    ii. Decision (plan to achieve a goal)
    iii. Connection
    iv. Consequence (consequences of performing the plan)
- Denouement

World facts are facts about the world that must be true in order that a theme structure can be applied; that is, they are preconditions. Decision represents a plan to achieve a goal. Consequence represents the consequences of performing the plan. Connections are events employed to make the transition from the Decision to the Consequence.

The Introduction and the Denouement are routines with the purpose of improving the final presentation of the story. The Introduction starts a tale by describing a time frame (a random year in the Middle Ages) and locating the main character at Camelot. The Denouement provides solutions to unsolved character goals and describes the fortune of characters. For instance, if after finishing the instantiation of the theme structure the life of a character is still at risk, this routine inserts an event that describes how that character escaped the dangerous situation; when a character dies, it creates a scene where the character is buried.

To develop a new story MINSTREL performs the following main processes:

1. Select and illustrate a theme structure. Once a theme has been selected it is necessary to satisfy the author goal of illustrating the theme. Themes are formed from a set of schemas representing incomplete scenes: one, two, or even all the attributes in each of the structures associated with the theme might not have an assigned value. Thus, the goal of illustrating a theme consists of employing those partially filled schemas as a search specification or index to retrieve scenes from episodic memory and in this way instantiate the schemas that make up the theme.

2. Generate the introduction and denouement of the story.

3. Link the story schemas to templates to produce the final output. To illustrate how the generation of English works, you can picture a character schema that includes the attributes role and name, which have the values 'knight' and 'Lancelot'. This structure is linked to the template 'a ?role named ?name' where ?role and ?name represent variables. In this case, the variables are substituted with 'knight' and 'Lancelot', respectively, producing the phrase 'a knight named Lancelot'. The same process is applied to the rest of the story schemas. The system includes routines for pronouns, capitalized letters, and so on.

As part of the process of developing a new story, each time a new scene is created the system tests whether an opportunistic goal can be triggered. This type of goal is used: (a) to verify the consistency of the story in progress; for example, when the system attempts to use a theme structure but its preconditions are not satisfied, a consistency opportunistic goal is triggered in order for MINSTREL to insert events that fulfil these requirements; and (b) to find opportunities to include dramatic elements in the story, such as suspense. In this work suspense consists of prolonging the resolution of an uncertain outcome; to use this dramatic element the system verifies whether the life of a character is at risk and whether the goal of the involved character is important for illustrating the theme. If these conditions are fulfilled the program might include circumstances like 'Lancelot tried to run away but failed' to produce suspense.

7.3.2 Creativity in MINSTREL: TRAMS

MINSTREL's main contribution is its representation of the creative process. Writing consists of instantiating all the schemas that make up the theme. When MINSTREL cannot find events in episodic memory to instantiate the theme, or all available events have been employed twice in previous stories (in MINSTREL, an episode that has been used more than once in a story does not satisfy novelty requirements), a set of heuristics called transform recall adapt methods (TRAMS) is employed to create novel scenes.

They work in the following way: TRAMs have explicit instructions on how to make small modifications to the schema's specifications used as indexes to recall items from episodic memory (i.e. how to modify the value of one or more of the schema's attributes). Those instructions are executed, creating a new slightly different index. MINSTREL searches memory to find an event which matches the new specifications. If no scene is found, a new TRAM is selected, producing new slightly different specifications, and a new attempt to match schemas in memory is launched. This process continues until an event is matched or a processing limit is reached. Once an event is matched, MINSTREL goes back step by step, adapting the event found to the original specifications in each step, until it finally reaches the top level.

The following example illustrates how TRAMS work. The goal is to create a scene where a knight kills himself. The episodes in memory are: (1) a knight fights and kills a troll and (2) a princess makes herself intentionally ill by drinking a potion. One of the solutions MINSTREL comes across when solving the problem is that a knight intentionally drinks a potion in order to kill himself. To achieve this result MINSTREL employs two TRAMS as follows.

The system selects TRAM-similar-outcomes-partial-change, which 'recognizes that being killed is similar to being injured . . . [that is, it] "guesses" that an action which is known to result in injury might also result in death' (Turner 1993, 116). Thus, the TRAM substitutes for the original description 'a knight kills himself' the new description 'a knight makes himself ill'. MINSTREL looks in memory for an episode that fulfils this condition and it fails.

A second TRAM is applied. TRAM-generalize-constraint substitutes the character 'knight' with 'anyone'; in this way, the new specification is 'someone does something to injure themself'.

This time the episode where the princess makes herself ill is recalled. The system returns step by step and princess is substituted with knight and a potion to make someone ill with a potion to kill someone. This produces the scene where the knight kills himself by drinking the potion.

7.3.3 Discussion about MINSTREL

Here we describe the strengths and limitations of MINSTREL (for an extended description and analysis of MINSTREL see Pérez y Pérez and Sharples 2004). This is a powerful and complex system capable of producing interesting and novel outputs. Its main features are the following:

- MINSTREL's main contribution to computational creativity in writing is the concept of TRAMs, which demonstrate the potential power of small changes in story schemas;
- MINSTREL explicitly employs author goals combined with character goals to generate adequate and interesting outputs; and
- MINSTREL generates novel stories by transforming 'old' episodes stored in memory.

Although TRAMs are a powerful tool, they sometimes appear to be written for the special purpose of achieving a specific scene. For example, in the case of the knight who wants to commit suicide, it is difficult to picture how TRAM-similar-outcomes-partial-change, which 'guesses' or 'recognizes that being killed is similar to being injured' (Turner 1993, 116), can work in a different context. To illustrate this situation, you can imagine a knight who is sewing his socks and pricks himself by accident; in this case, because the action of sewing produced an injury in the knight, MINSTREL would treat sewing as a method for killing someone. Something similar occurs with TRAM-similar-outcomes-implicit, another heuristic for interchanging story elements. This heuristic suggests that 'two outcomes are interchangeable in every situation if it [the system] can recall any situation in which they are interchangeable' (Turner 1993, 120). Thus, if in episodic memory there is a scene where a knight kills a monster, and another scene where a knight kills a dragon, MINSTREL 'guesses' that monsters and dragons are interchangeable. One wonders what happens if in memory there is a scene where a knight kisses his horse after a battle, and in the following scene the knight kisses his girlfriend. Would MINSTREL 'guess' that girlfriends and horses are interchangeable? Turner is aware of the possibility of producing strange outputs and explains it by saying that 'A creative problem solver should make errors' (Turner 1993, 120). MINSTREL would benefit from incorporating a module that makes it possible to learn from such mistakes.

One of the most impressive characteristics of MINSTREL is its capacity to create stories where revenge, deception, mistaken beliefs, and so on take place. However, they have all been previously programmed in ten heuristics that explicitly indicate the structure that such scenes have and the way to construct them. Similarly, MINSTREL can only produce stories with six different themes, which are structurally predefined and which only can turn around a planning process. These features limit its flexibility.

MINSTREL is a very complex program which has pointed out the utility of small modifications as a way of solving problems. Although TRAMs in MINSTREL seem at times to be too specific, they are a powerful tool during problem solving. Also, MIN-STREL indicates the importance of author goals in storytelling, particularly theme, consistency, drama, and presentation goals. The most important limitations of MIN-STREL are TRAMS tailored to the specific genre of a story, inflexibility in a story's structure, and rigidity in the structure of some scenes.

7.4 Final remarks about planning

Storytellers based on planning and problem-solving techniques are complex systems. Their knowledge bases include significant amounts of actions, goals, and themes. Computers need to probe all valid combinations of these elements in order to generate diverse narratives. Beside this, when trying to achieve a story goal, the program performs a set of story actions and/or other story goals, which in turn might include new story actions and story goals to be performed, and so on. Furthermore, during this process the system may choose options whose preconditions cannot be fulfilled; as a result, it needs to go back and undo some of the deeds, then look for new alternatives. On top of this, storytellers need to manage the instantiation of theme structures, supervise the production of suspense, flashback, and so on, and operate the storage, retrieval, and adaptation of scenes from episodic memory. For humans it is hard, maybe impossible, to satisfactorily complete all these tasks in order to generate a narrative; a computer can produce appealing outputs in just seconds. These types of programs are able to represent abstract concepts like story themes, characters' needs, fears, and dreams, authors' plans and aims, and narrative resources like flashback and suspense. They are all essential concepts for understanding automatic narrative generation. Their main limitation is that those actions, goals, themes, opportunistic goals, and the like, as well as the mechanisms to manage them, need to be defined in advance by the designer. Thus, a storyteller will always write tales about those themes that the designer has stored in its knowledge base.

8
Weaving texts with patterns

The power of repetition

8.1 Introduction to statistical information techniques

Building an automated narrator based on goals and plans is necessary, but not suffi-
cient. The narrator must also tell the story in appropriate and appealing language.
Natural language processing (NLP) is the study of computational models to process language (Wilson and Keil 1999). This research area focuses on two core topics,
comprehension and generation of language. Statistical natural language processing
is the development of computer programs that perform language-related tasks based
on statistical information collected in large corpora (structured collections of texts).
This chapter describes how NLP statistical information techniques can be employed
to generate texts.

Think of your favourite novel. We can easily develop a computer program that lists,
for any given word in the novel (e.g. 'love', 'automobile', or any other that you prefer),
which words appear next in the text. In addition to obtaining this group of 'succeeding words', the system can also calculate the number of occasions that each word
appears in this group. Thus, it is possible to estimate its probability of occurrence.
Here are a couple of examples to illustrate these ideas.

Suppose the word 'love' occurs only once in the entire novel, say, in the sentence
'Love disappeared'. In this case, because there is only one possible option, 'disappeared' has a 100% chance of continuing after 'love'. Now, let us say that the word
'love' occurs five times in the text: 'love disappeared', 'love hurt', 'love evolved', 'love
disappeared', and 'love disappeared'. One of the phrases ('love disappeared') occurs
three times in the book. In this context, our computer system detects that the group
of words that succeed 'love' are 'disappeared' with three occurrences, 'hurt' with one,
and 'evolved' also with one. You can picture this example as a box that is labelled on
the top with the string 'love'. Inside it the reader puts three pieces of paper with the
word 'disappeared', one piece of paper with the word 'hurts', and one with 'evolves'
(see Figure 8.1). Thus, in our favourite novel, 'disappeared' has a 60% chance of
appearing after 'love', while 'hurts' and 'evolves' have 20%, respectively. The set of
all the words that follow 'love', and their probabilities of occurrence, is known as the
probability distribution. This concept plays a vital role in this type of system. In this
chapter we use the term *reference* to refer to the word for which we want to know the
group of words that succeed it. In the previous example, the reference is 'love'.

An Introduction to Narrative Generators. Rafael Pérez y Pérez and Mike Sharples, Oxford University Press.
© Rafael Pérez y Pérez and Mike Sharples (2023). DOI: 10.1093/oso/9780198876601.003.0008

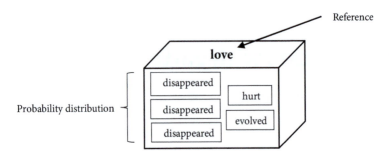

Figure 8.1 Representation of the probability distribution to continue the word 'love'

Next, let us imagine we develop a program capable of generating the probability distribution of each word that makes up our novel. The number of elements to consider would be high; this number depends on the size of the novel and the writing style of the author. The most used words will produce huge groups of possible successors while those hardly used words will generate very small groups. Because most texts include not only words but also elements like punctuation marks, it is also necessary to take those components into consideration. Let us develop an example that allows for deeper exploration of these ideas.

8.2 'The Devoted Friend'

For this exercise we use a short story by Oscar Wilde entitled '*The Devoted Friend*' that you can download from the Gutenberg Project.[1] The tale describes the abusive relationship that Miller, a wealthy man, has with little Hans, a modest gardener who is in love with his flowers. The constant abuse by Miller, and the need for little Hans to be considered a true friend of this man, triggers a situation that ends tragically. This piece comprises 4334 words. Table 8.1 shows the probability distribution of some of these words that we will employ in the following examples.

The column Reference (Ref) shows the word used as a reference; Occurrences (Occ) indicates the number of times that this reference word appears in the text; Number (Num) shows the number of different words that follow the reference; the last column shows the reference's probabilistic distribution of the group of possible successors. Each word in this group is accompanied to the right by its probability of occurrence based on the information obtained from Wilde's tale. For instance, the word 'best' appears seven times in the text and is followed by three words: 'friend' which has 71.4% probability of succeeding the reference, and 'society' and 'place' which each have 14.3% probability. The group of possible successors for the word 'Miller' includes punctuation marks such as commas, semicolons, periods, possessive

[1] https://www.gutenberg.org/files/902/902-h/902-h.htm accessed on 26 August 2020.

Table 8.1 Probability distributions of thirteen elements from 'The Devoted Friend' by Oscar Wilde

Ref	Occ	Num	Probabilistic distribution of the group of possible successors
all	25	17	the (16.0%) my (16.0%) kinds (8.0%) over (4.0%) . . .
best	7	3	friend (71.4%) society (14.3%) place (14.3%)
cottage	3	3	all (33.3%) , (33.3%) . (33.3%)
cried	14	4	little (42.9%) the (42.9%) Hans (7.1%) his (7.1%)
is	68	34	the (13.2%) a (8.8%) not (5.9%) quite (5.9%) very (4.4%) just (4.4%) no (2.9%) in (2.9%) another (1.5%) generous (1.5%) all (1.5%) . . .
love	1	1	is (100.0%)
Miller	54	17	, (35.2%) 's (11.1%) ; (9.3%) . (5.6%) used (5.6%) was (5.6%) never (3.7%) said (3.7%) with (3.7%) came (3.7%) to (1.9%) immensely (1.9%) tied (1.9%) in (1.9%) brought (1.9%) wants (1.9%) ? (1.9%)
never	13	13	able (7.7%) be (7.7%) been (7.7%) came (7.7%) forgets (7.7%) gave (7.7%) go (7.7%) have (7.7%) intend (7.7%) look (7.7%) mind (7.7%) notices (7.7%) troubled (7.7%)
over	7	5	the (28.6%) , (28.6) him (14.3%) with (14.3%) at (14.3%)
society	2	2	unless (50.0%) at (50.0%)
the	258	117	Miller (20.2%) Water-rat (5.0%) Linnet (4.3%) winter (2.3%) best (0.8%) cottage (0.8%) . . .
unless	1	1	you (100.0%)
's	11	10	house (18.2%) place (9.1%) youngest (9.1%) nature (9.1%) Wife (9.1%) daughter (9.1%) basket (9.1%) voice (9.1%) son (9.1%) feelings (9.1%)

apostrophes, and question marks. The last row in Table 8.1 shows the probability distribution of the possessive apostrophe ('s) in the story. Finally, for the sake of clarity, Table 8.1 does not include all the members of the group of possible successors for the words 'all', 'is', and 'the'.

8.2.1 Let us generate some text!

We can employ the information in Table 8.1 to develop a system that generates novel sentences five words long. It works as follows. The user chooses an initial word as a reference. The system looks for it in the first column of Table 8.1 and selects one of the options inside the group of possible successors. The selected item is displayed as the next word in the sentence and therefore it substitutes for the previous reference. The same process starts all over again until the sentence is finished.

The following illustrates this process. Let us start with the familiar 'love' as the initial reference. Table 8.1 indicates that the only option to continue the sentence is the word 'is' which is now used as reference (it is marked in bold):

'love' → '**is**' (100.0%)

So far, our system has produced 'love is'. Next, Table 8.1 shows that 'is' has thirty-four possible options to continue. The system can choose one of them at random, or it can pick the word with the highest probability, or the word with the lowest probability, and so on. For this exercise, the program chooses at random between those words with the two highest probabilities. In this case, the options are 'the' and 'a'; let us assume that the system randomly chooses the first one as the next word:

'is' → **'the'** (13.2%), 'a' (8.8%)

Thus, so far, we have produced 'love is the'. Next, the system looks in Table 8.1 for the reference 'the' and finds that 'Miller' and 'Water-rat' are the two options with the highest probability; it selects at random the first one:

'the' → **'Miller'** (20.2%), 'Water-rat' (5.0%)

At this point the system has generated 'love is the Miller'. The system verifies the options for the reference 'Miller' and finds a coma ',', and a possessive apostrophe ''s' as possibilities; it selects the second option:

'Miller' → '', (35.2%), **''s'** (11.1%)

The system generates 'love is the Miller's'. Table 8.1 shows that the most popular words that follow the reference ''s' are 'house' and 'place'; it selects the first one and the system generates 'love is the Miller's house'.

''s' → **'house'** (18.2%), 'place' (9.1%)

The system has already produced a sentence with five words and therefore the process ends. In this way, starting with the word 'love', the system is able to generate 'Love is the Miller's house' (might this be considered a poetic way to express that people living at the Miller's house love each other?). This is a novel sentence that cannot be found in the original story.

This example is useful to point out relevant features of this system.

(a) *Novel material.* If we analyse the generated text as pairs of words, we observe that 'Love is' occurs one time in the original story, 'is the' appears nine times, 'the Miller's' six times, and 'Miller's house' once. Because of the way the program works, it makes sense that we always find in the original story at least one occurrence of any pair of words generated by the system. However, if we analyse the output as sequences of three words, then we realize that 'love is the' and 'is the Miller's' are not present in Wilde's tale:

Love is the Miller's house
Love **is the Miller's** house

Oscar Wilde wrote in his tale: 'In fact, I have never been married, and I never intend to be. *Love is all* very well in its way, but friendship is much higher' (italics are ours). Because 'Love is' appears only once in the whole text and it is followed by 'all', any other word that succeeds this pair will generate a new sentence. In other words, as shown in Table 8.1, from thirty-four options that the system can choose to follow 'love' + 'is', thirty-three result in a sequence not found in the original text. Thus, novelty arises when the system is able to join unseen series of words, resulting in sentences not included in the original text. The following line shows another example of a novel output from all the words in the story (apart from the first two, the words in this example are not found in Table 8.1). It has a length of eleven elements and during the generation process the system chooses at random the next item to progress the text between the five options with the highest probability:

Love is a very dangerous place; so he is.

(b) *Generation of coherent sentences.* Because the system (re)uses the story written by Wilde to produce its outputs, we are certain that all pairs of words that our program generates make sense; that is, the system never will produce sequences like 'the !' unless they are present by mistake in the original text. However, the program often fails to generate coherent outputs. This problem increases as the length of the output grows. That is because, in the generative process that we have described, there is only a probabilistic relationship between each element of the sentence and the next. To describe it in a metaphorical way, the current word in the sentence only 'talks' to the previous one and has no 'knowledge of the existence' of any element prior to it. Thus, as the length of the output increases, it may become less coherent. The following line illustrates this situation; it has a length of twenty-two elements and during the generation process the system chooses at random the next item to progress the text between the two options with the highest probability:

Love is a great loss to give you are certainly a great many friends should have given him.

(c) *Repetition of the same sentence.* The system can get trapped in a loop resulting in constant repetition of the same sequence of words. The following line illustrates this situation:

Said the Water-rat, said little Hans, said the Water-rat . . .

For this example, we employ 'Said' as the initial reference. During generation, the system chooses at random the next item to progress the text between the two options with the highest probability. With the help of Table 8.2 we can observe the path that the program follows to get trapped in the loop. 'Said' is followed by either 'the' or 'little'. These options have a 50% probability of choosing either 'Hans' or 'Water-rat', both of which have a 50% change of selecting a comma to progress the text, which in

Table 8.2 Probability distributions of five words from 'The Devoted Friend' by Oscar Wilde

Ref	Occ	Num	Probabilistic distribution of the group of possible successors
Hans	61	23	, (41.0%) was (9.8%)…
little	52	8	Hans (84.6%) ducks (3.8%)…
said	48	8	the (66.75) little (12.5%)…
the	258	117	Miller (20.2%) Water-rat (5.0%)…
Water-rat	15	5	. (60.0%) , (13.3%)…
,	390	72	and (33.6%) said (8.7%)…

turn has 50% of probability of coming back to 'said', the starting point. Because the first and the last word are the same, the system can get trapped in a loop repeating the same sequence of words. To avoid this problem, during generation we can expand the number of available options to select the next item in the sequence. This reduces the possibilities of getting trapped in a loop. However, it also increases the likelihood it produces incoherent outputs.

The following summarizes how this system works:

1. The user provides an initial word that we call *reference*.
2. The system obtains the probability distribution of the words that follow the reference.
3. The system randomly chooses one word among those options with the highest probability of occurrence. The selected word is appended to the end of the sentence that the system is generating.
4. The word selected in step 3 becomes the new reference; the system returns to execute step 2.

This cycle is repeated until the system generates a text that satisfies a given criteria (e.g. length of the output).

8.3 Improving the system with patterns

A word is a series of letters. Some combinations of letters form patterns that are recognized as concepts, such as 'cat' or 'dog'. Texts are made up of a diversity of patterns. Expressions like 'hold on', 'look at', and 'fond of' illustrate the recurring use of pairs of words. Some familiar sequences include a larger number of items, like 'is able to', 'looking forward to'. Idioms are examples of patterns in English, such as 'a drop in the ocean', 'a bitter pill', 'bits and pieces', 'over and over again', 'cut no ice', 'behind closed doors', 'from cover to cover', and 'on the house'. Pattern recognition is a useful process because identifying and manipulating patterns to use them as single items during the generation of a text (i.e. as if they were one word) can help our system to produce more elaborated outputs. If a system can identify one or more patterns that a writer frequently employs in diverse works, then it is possible to generate a text that reproduces those or similar features. Thus, the system would be able to mimic part of the style of that writer. Similarly, authors may devise characters who regularly utter the same phrase to stamp part of their personality. For example, imagine a story-character that frequently ends his statements during a dialogue with 'Do you know what I mean?' A system might reproduce this kind of distinctive mannerism.

Let us delve into these ideas. For this exercise, we establish that a sequence of words is categorized as a pattern when it appears at least three times in 'The Devoted Friend'; for example, the sequence 'you my wheelbarrow' repeats nine times:

- I will give **you my wheelbarrow**. It is not in very good repair;
- Yes, you may set your mind at ease, I will give **you my wheelbarrow**.
- I have given **you my wheelbarrow**, and now you are going to give me your plank.
- And now, as I have given **you my wheelbarrow**, I am sure you would like to give me some flowers in return.
- as I have given **you my wheelbarrow**, I don't think that it is much to ask you for a few flowers.
- I think that, considering that I am going to give **you my wheelbarrow**, it is rather unfriendly of you to refuse.
- Really, considering that I am going to give **you my wheelbarrow**, I think you might work harder.
- I do not think it is much to ask of you, considering that I am going to give **you my wheelbarrow**;
- You know I am going to give **you my wheelbarrow**, and so, it is only fair that you should do something for me in return.

If we consider this pattern as a single item, we can easily obtain information like that shown in Table 8.1. So, 'you my wheelbarrow' is followed two times by a period, six times by a coma and one time by a semicolon:

'you my wheelbarrow' → '.' (22.2%), ',' (66.7%), ';' (11.1%)

If we pay closer attention to the above nine sentences, it is possible to discover at least two more patterns:

'I am going to give you my wheelbarrow'
'I have given you my wheelbarrow'

The first sequence happens four times in the tale and the second occurs three times. And we can keep going on. The sequence 'I am going to' occurs six times in the tale and is followed four times by the pattern 'give you my wheelbarrow', and once by the words 'bring' and 'buy':

'I am going to' → 'give you my wheelbarrow' (66.6%), 'bring' (16.7%), 'buy' (16.7%)

The pattern 'I am going to' is also part of the longer pattern 'I am going to give you my wheelbarrow'. Thus, two or more repetitive sequences of words can be used to build a more elaborated expression.

Patterns that include punctuation marks also provide useful information. For instance, spoken dialogues usually are indicated by quotation marks or double angle quotation marks. Thus, the system can identify how to continue a sentence when a character asks a question. In 'The Devoted Friend', the sequence indicating the end of a question in a dialogue between characters ('?»') is followed by 'asked' seven times, 'said' and the symbol to initiate a new dialog ('«') four times, 'cried' two times, and 'he' and 'screamed' one time.

'?»' → 'asked' (37%), 'said' (21%), '«' (21%), 'cried' (11%), 'he' (5%), 'screamed' (5%)

So, if the input to the system is '«hello?»', employing this information together with that in Table 8.1, the system might generate '«hello?» cried the Miller'. This illustrates how patterns encode data that help to identify distinctive situations within a text, also known as contexts; in this way, identifying a sequence that represents that a character has just asked a question makes it possible to generate more elaborated and well-structured paragraphs. Similarly, because 'An idiom can be defined as a number of words which, when taken together, have a different meaning from the individual meanings of each word' (Seidl and McMordies 1991, 13), the use of patterns allows the system to generate texts that humans can interpret as expressing particularized semantic content.

8.4 Taking advantage of digitized data

Using a single story causes serious limitations to our system. Given that hundreds of thousands of digital texts are available today, the next step is to employ a much greater amount of writing. That is, instead of just using 'The Devoted Friend' as a dataset, we might employ the complete collection of short stories by Oscar Wilde; or better yet, all his work. If we want to be more ambitious, we can incorporate the work of other contemporaneous authors, or the books from writers from other latitudes or other times. We can even think of including texts from blogs, social networks, newspapers, and so on. As a result, the system will have at every step the opportunity to choose between an enormous number of words and patterns. In this way, it is possible to build more sophisticated outputs. Similarly, these attributes make it possible to perform powerful processes like *generalization*.

Generalization consists of identifying from a set of input patterns those features that are significant. In this way, new patterns can be associated with others that share the same distinguishing characteristics (Beale and Jackson 1992, 89). Let us elaborate on this idea. If the system's dataset includes enough texts, it is possible to detect coincidences and dissimilarities between hundreds of different repetitive sequences of words to create more general patterns. For instance, let us imagine that the system works with a dataset where the sequence 'That morning Jenny went to the supermarket' repeats 300 times, the sequence 'That morning Tom went to the church' repeats 180 times, and 'That morning I went to the park' repeats 450 times

That morning Jenny **went to the** supermarket
That morning Tom **went to the** church
That morning I **went to the** park

The system detects that 'That morning ___ went to the ___' repeats in all cases, producing a new pattern with blanks. (This sounds like the templates that we studied in Chapter 2!) The possible options to continue this pattern are determined by the content of the dataset. In this way, if the program has no information about Victor Hugo's *Les Misérables*, it should be capable of progressing the sentence 'That morning inspector Javert went to the bridge' by identifying the pattern 'That morning ___ went to the ___' and employing one of the available options to progress the string. If the dataset is large enough to include several detective novels, the system might even identify variations of the same pattern, for example, 'That morning ___ went to the ___' and 'That morning Inspector ___ went to the ___'. The former would be more general than the latter; the latter would be mainly associated with activities that inspectors usually perform. That is, it includes more contextual information. Patterns can be generalized to include gaps for additional words—but with the danger that the inserted words may well be irrelevant to the story. So, because of its capacity to identify slightly different patterns, the program can correctly classify expressions like 'That morning Inspector Javert went in a hurry to the bridge'. In this way, generalization allows the system to manipulate texts that it 'has never seen before', that is, texts not included in its dataset.

8.5 An example is worth a thousand words

In this exercise we progress a sentence using those patterns and words shown in Table 8.3. We start the program with the input 'The little boy' and the system generates as output 'The little boy and he ran into my best friend'. Although this system works in a similar way to those described previously such as in Section 8.2, it requires some adjustments to handle patterns. So, in this new version:

1. In addition to the probability distribution of words, the system obtains the probability distribution of patterns in the dataset.

Table 8.3 Probability distributions of some words and patterns from 'The Devoted Friend'

Ref	Occ	Num	Probabilistic distribution of the group of possible successors (words and patterns)
Hans, and he	3	3	began (33.3%) ran (33.3%) took (33.3%)
		1	began to (100.0%)
his best friend	3	2	, (66.7%) . (33.3%)
		3	, and I (33.3%) , and I will (33.3%) , said the Miller (33.3%)
little Hans,	17	11	and (23.5%) said (17.6%) I (11.8%) . . .
		8	and he (23.1%) said the Miller (23.1%) said the Miller, (15.4%) cried the Miller (7.7%) and he ran (7.7%) . . .
into	7	4	the (42.9%) his (28.6%) my (14.3%) any (14.3%)
		0	——
my	44	27	wheelbarrow (27.3%) barn-roof (4.5%) best (4.5%) . . .
		7	wheelbarrow, and (23.1%) wheelbarrow, I (23.1%) best friend (15.4%) best friend, (15.4%) silver buttons (7.7%) . . .
ran	5	5	and (20.0%) down (20.0%) in (20.0%) into (20.0%) to (20.0%)
		0	——

2. Previously, the reference always was the last element in the text in progress. Now, the reference includes up to the last five items of the text in progress (this number can be modified).

3. The generation process starts when the user provides an input that works as the initial reference; for this exercise, the input might be up to five words (previously, it included only one word).

4. The system looks for the reference in the first column of Table 8.3 and selects one of the n-best options inside the group of possible successors (the best options are those with the highest probabilities). When the system has the alternative to choose whether the reference is proceeded by a word or by a pattern, it gives preference to the latter. This helps to produce more elaborated outputs. If there are not enough number of patterns to complete the n-best options of possible successors, the system might add the best words until the required quantity is reached. The selected item is the next element in the sentence and therefore it is appended to the output. The same process starts all over again until the sentence is finished.

For this new version, when the reference is a pattern, we need to pay particular attention to the part of the process where the system selects how to progress a sentence. It works as follows:

i. The system looks in the first column of Table 8.3 for an entry that is equal to the reference.

ii. If the previous step fails, then the system looks in the first column of Table 8.3 for an entry that is similar to the reference. For this exercise, 'similar' means that at least 50% of the elements in the reference can be found in the entry or vice versa; this criterium might be modified according to the purposes of the program. Words like 'a', 'an', and 'the' and punctuations marks are not considered, so only the relevant elements of each sequence are compared. Matching patterns that are similar allows the program to generalize.

iii. If none of the previous steps work, then the system looks in the first column of Table 8.3 for an entry that is equal to the last word of the reference. Thus, a reference might be either a word or a pattern (previously, the reference was only a word).

The generation process ends when the output reaches a length of 6 items, that is, a length of six words and/or patterns (this is an arbitrary parameter that can be modified).

8.5.1 Details of the dataset

Table 8.3 shows the set of words and, in the next line, the set of patterns that follow each reference employed in this example (the whole dataset is much bigger). The dataset only comprises text from 'The Devoted Friend', to easily track the generating process. For instance, the sequence 'little Hans' repeats 17 times in the tale. Here are some samples:

Indeed, so devoted was the rich Miller to **little Hans,** that he would never go . . .
«Quite full?» said **little Hans,** rather sorrowfully, for it was really a very big basket . . .
«I am very sorry,» said **little Hans,** rubbing his eyes and pulling off his night-cap . . .

Clearly, 'little Hans' is one of the patterns that can be obtained from the data; it has three elements: two words and one comma. As shown in Table 8.3, this sequence is followed by eleven words and by eight patterns. The following shows another case:

I am **his best friend,** and I will always watch over . . .
. . . but he consoled himself by the reflection that the Miller was **his best friend.**
«As I was **his best friend,**» said the Miller . . .

The sequence 'his best friend' includes three words, repeats three times and, as Table 8.3 indicates, it is followed by two words and three patterns. The rest of the information in Table 8.3 is obtained in a similar way.

8.5.2 The little boy

Let us start this exercise with the sequence 'The little boy'. The system checks the first column of Table 8.3 and finds the pattern 'little Hans'. The initial reference and the pattern in Table 8.3 are not identical, so the program checks if they are similar. To compare them, it eliminates from the reference the word 'the' and from the sequence in Table 8.3 the comma. This results in the strings '**little** boy' and '**little** Hans'. Because they are 50% similar, they are matched.

Table 8.3 shows that the sequence 'little Hans', is followed by eleven words and eight patterns. The system gives preferences to the patterns:

little Hans, → **and he** (23.1%) said the Miller (23.1%) said the Miller, (15.4%)
cried the Miller (7.7%) and he ran (7.7%)

The program chooses at random the first option. So far, the generated output looks as 'The little boy and he' and the reference is set to the last five words of the output 'The little boy and he' (in this case, they are the entire output).

Next, the system looks for an option to match the reference. It finds in the first column of Table 8.2 the sequence 'Hans, and he', which satisfies our criteria for similarity because 'little boy **and he**' and 'Hans **and he**' are at least 50% similar. This entry is followed by one pattern and three words:

Hans, and he → began to (100.0%) began (33.3%) **ran** (33.3%) took (33.3%)

The system selects at random 'ran'. So far, the output looks as 'The little boy and he ran' and the refence is set to 'little boy and he ran'. This time the reference cannot match any pattern, so the program looks in Table 8.3 for 'ran', the last element in the reference. This is followed by five words and zero patterns:

ran → and (20.0%) down (20.0%) in (20.0%) **into** (20.0%) to (20.0%)

The system chooses at random 'into'. The current output is 'The little boy and he ran into' and the reference is set to 'boy and he ran into'. Again, the system cannot match any pattern and looks for 'into' in Table 8.3, which is followed by four words and zero patterns:

into → the (42.9%) his (28.6%) **my** (14.3%) any (14.3%)

The system chooses at random 'my'. The current output is 'The little boy and he ran into my' and the reference is set to 'and he ran into my'. Again, the system cannot match any pattern and looks for 'my' in Table 8.3:

my → wheelbarrow, and (23.1%) wheelbarrow, I (23.1%) best friend (15.4%)
best friend, (15.4%) silver buttons (7.7%)

The option 'best friend,' is selected at random. At this point the system has generated a six-item output 'The little boy / and he / ran / into / my / best friend,' and it stops. This sequence is novel because 'The Devoted Friend' does not include this sentence.

Thus, the system we have described works as follows:

1. The designers of the system gather a collection of enough texts; we refer to these as a *dataset*.
2. The computer system processes this dataset to calculate the probabilistic distributions of words and patterns.
3. The user of the system provides an input, namely, a sequence of words.
4. The system progresses the input until the condition to stop the process is reached, such as when the output reaches a predetermined length.

8.6 Refining our probabilistic methods for narrative generation

The methods that we have studied through this chapter represent, in terms of probabilistic distributions, the knowledge to progress a narrative. The designer of the system must collect diverse texts to build the dataset. The quality of this corpus is important because an output's core features like the vocabulary and the content strongly depend on the attributes of the dataset.

There are different criteria that can be employed to select the next element to progress a narrative. In the examples that we have presented in this chapter, we chose those items with high probabilistic distribution. Words and patterns with high probabilistic distribution are more common in the dataset than those with low probabilistic distribution. This information is useful for generating narratives that attempt to reproduce common elements found in the original corpus of texts; similarly, they can be used to avoid reproducing those elements. Big datasets that collect information from different sources, like blogs or social networks, are susceptible to include erroneous data, such as orthographical errors or incomplete words. In this case, high probability distributions can work as filters that help to avoid using mistaken elements (usually erroneous data have low probabilistic distributions).

8.6.1 Polishing the final product

It is common to develop routines that improve the text generated by such a program. These might include procedures that eliminate elements that repeat themselves in close succession; check that sentences always end with a period; check that two pronouns are separated by commas, and so on. Simple changes like those might considerably enhance a sentence. We can develop even more powerful routines that check, at least for some cases, the consistency of the tense of verbs, verify the correct use of pronouns, and so on. Although it is unfeasible to cover all the possible alternatives for enriching a text, a carefully chosen set of these types of tools might be effective in producing enjoyable outputs. However, correcting grammatical mistakes does not ensure that the sentence makes sense. More important, it does not produce a coherent narrative.

8.6.2 Overcoming some limitations

The technique we have been studying throughout this chapter is known as a Markov Chain (Basharina et al. 2004). As we mentioned earlier, programs that employ this technique do not consider what occurred earlier in the text when it generates the next sentence. They can be described as systems with 'no memory'. This is an important limitation because the way events unfold in a narrative constrain the possibilities of future incidents. In other words, plots show a strong dependency between what happens earlier and what happens next. In this section, we briefly discuss some procedures that help to identify lengthier probabilistic relations between elements in a text. These procedures contribute to reducing the 'no memory' limitation described and offer new possibilities for the generation of narratives.

The sequence 'Little Hans was' in 'The Devoted Friend' occurs six times, five of which are followed within the next couple of sentences by the sequence 'he was'. These are some examples:

> Early the next morning the Miller came down to get the money for his sack of flour, but **little Hans was** so tired that **he was** still in bed.

> Poor **little Hans was** very anxious to go and work in his garden, for his flowers had not been watered for two days, but he did not like to refuse the Miller, as **he was** such a good friend to him.

> **Little Hans was** very much distressed at times, as **he was** afraid his flowers would think he had forgotten them, but he consoled himself by the reflection that the Miller was his best friend.

There is a clear relation between 'Little Hans was' and 'he was'. We refer to this type of association as a *long-distance association* because it is a structure that includes two related elements separated by an unknown number of items. The maximum number of allowed unknown items that separate the two related elements is defined by the designer of the system.

Long-distance associations provide information about the text that it is hard to obtain in other ways. For this example, practically each time that Oscar Wilde provides a description about how 'little Hans was', the author also includes a brief complementary description of the current situation; for example, 'little Hans was tired' is complemented with the fact that 'he was in bed', 'little Hans was anxious' is complemented with the fact that 'he was a good friend', 'little Hans was distressed' is complemented with the facts that 'he was afraid'. Thus, if one wants to develop a computer program that generates narratives in the style of 'The Devoted Friend' the system needs to be able to reproduce this type of recurrent string.

Long-distance associations can be pictured as patterns with a variable number of elements between their fixed parts. They can be represented visually as 'fixed-part-1 *** fixed-part-2' where the stars symbolize an undetermined number of elements between the two fixed parts; for example, 'Little Hans was *** he was'. A dataset might include more elaborated long-distance associations with multiple fixed and variable parts. Also, it is possible to build structures that establish relations between two or more long-distance associations. Structures that include as elements words, patterns, and other long-distance associations resemble the templates we studied in Chapter 2.

Long-distance associations are useful for identifying relevant features that characterize a text and that are not recognized by systems based exclusively on a Markov Chain technique. This information can be exploited in diverse manners, such as supporting the generation of writings in a particular style, and supporting the production of coherent outputs by reproducing narrative structure features found in the original texts. However, it is impossible to know in advance all the words, patterns, and long-distance associations that a dataset might include. Thus, the process to automatically identify long-distance associations is complex and, depending on the size of the dataset, it might require considerable computer power. Similarly, the use of long-distance associations during the generation of a narrative requires the development of specialized routines that can manage the unknown elements.

8.7 *A Noise Such as a Man Might Make*

We have a little boy. What are we going to do? You have to drink he said. I'll give him the belly meat of a big fish, the female, made a wild, panic-stricken, despairing fight that soon exhausted her, and all the way to the coast. He didn't know sharks had such handsome, beautifully formed tails. I did. Lots of them? We do. Do you think I lie to you? I mustn't try for the head. I'm starving, man. I done what you said. Listen to the kid. You tried to kill us. They might try to.

We haven't eaten in a long time? I think if we cook them really good they'll be all right. How do you feel, hand? He asked himself. Un espuela de hueso. We don't know. But I can't. He was coughing every step of it. He kneeled against the bow, he could not slow the fish with a steady pressure, the fish could take out all the quart plastic oil bottles.

(Läufer 2018)

The previous paragraphs are the opening of *A Noise Such as a Man Might Make*, a 126-page novel generated by a computer program that employs Markov Chains and which was developed by Milton Läufer. The dataset is composed of two novels: *The Old Man and the Sea* by Ernest Hemingway, published in 1952, and *The Road* by Cormac McCarthy, published in 2006. These works share important features that contribute to their integration; for example, rather than describing a plot that moves forward both texts portray what Läufer calls a series of 'small rituals' that involve, respectively, a man, and a man and a boy, in an endless journey, all set in a hostile environment. Similarly, the two novelists employ some comparable writing styles; for example, in both cases proper names are seldom used.

In a personal communication (December 2020/January 2021) Läufer explained to us how his system works. The author combines both novels into a single text file that forms the dataset. Then the program calculates the probabilistic distribution of sequences of two, three, and four words. Then the author provides an initial set of words to trigger the generation process. In this case, Läufer chose 'We' and 'have' because they can be understood as representing either a possession or an obligation and therefore, they offer different possibilities for progressing the novel (the initial words can also be chosen at random). Next, the generation process starts. Most of the time the system employs tetragrams (sequences of four words) as a reference to choose the next item in the sentence, which is selected at random. This helps to produce coherent although unsurprising sentences. Occasionally, the system employs bigrams or trigrams (sequences of two or three words) to progress the narrative and, in this way, surprising but at times incoherent texts emerge. Punctuation marks are processed as words.

The entire process that Läufer and his program followed to generate the novel includes five steps:

1. The computer program generated five versions of the novel of around 55,000 words each.
2. The author selected the one that he liked the most.
3. The program eliminated repeated elements in the selected text.
4. The author deleted sentences that he disliked or were senseless. In this way, the two previous steps reduced the text by around 20%.
5. The program divided the resulting narrative into ten parts that contained a similar number of words. Those are the ten chapters that compose the novel.

A Noise Such as a Man Might Make illustrates the potential of probabilistic methods for narrative generation.

8.8 Final remarks about probabilistic methods for narrative generation

In summary, the main strengths of automatic narrative generators based on the probabilistic approach described in this chapter are as follows:

- It is simple to entirely change the dataset without modifying the program. Thus, one can develop a system that employs the whole collection of Oscar Wilde's works to produce new texts. Then, employing the same program, one can generate new texts but this time with a dataset that comprises Gabriel García

Márquez's writings. We can also combine these corpuses or build new ones. Once the program has been developed, one can easily experiment with diverse datasets.

- This type of systems makes it possible to recollect statistical information that can be exploited in new narratives. For instance, the analysis of the most used words, patterns, and long-distance associations in the corpus makes it possible to figure out features that can be reproduced during the generation process. In this way, the automatic narrator can generate texts that resemble a particular style.

Some of the main limitations of this type of approach are:

- The program has no mechanisms to determine the coherence of the text it is generating.
- Important narrative elements that we have studied in previous chapters, such as the representation of story structures, characters' goals, and protagonist–antagonist relationship, are not represented.
- This type of systems is memoryless. Thus, the context of the narrative is not considered during the generation process and the program tends to produce texts where elements are repeated.
- Core processes, like calculating the dataset's probabilistic distribution or identifying patterns and long-distance relations, might require considerable computer power, especially, for large datasets.
- The construction of reliable large datasets might require developing complex routines to collect the data and eliminate unuseful material.

In Chapter 10 we will discuss a probabilistic method called deep neural networks, which is able to identify multiple patterns and long-distance associations within enormous datasets and use them to generate texts. To explain how deep neural networks work, in Chapter 9 we first introduce the notion of artificial neurons.

9
Inspired by the brain

A single neuron

9.1 Introduction to networks of artificial neurons

For years, the brain has been an inspiration for the creation of artificial intelligent systems. Almost 80 years ago, McCulloch and Pitts (1943) presented a mathematical model inspired by the basic functioning of a neuron. Nineteen years later, Rosenblatt presented in his book *Principles of Neurodynamics* (Rosenblatt 1962) an improved conceptual model of a neuron that he called a 'perceptron'. Subsequently, Minsky and Papert (1969) illustrated the limitations of the perceptron that led to a significant loss of interest in the area. However, after the publication of the *Parallel Distributed Processing* book (Rumelhart & McClelland 1986), which posed solutions to many of the perceptron's problems, the field gained new impetus. Since then, the study of neural networks has been growing. In this chapter, we start our study of neural networks by considering the performing of a single artificial neuron.

Artificial neurons are computer data structures that simulate in basic ways the process of activating a human neuron. Deep neural networks (DNNs) are computer programs that represent a huge number, millions or even billions, of interconnected artificial neurons. Neural networks are mathematical models. The function that describes the behaviour of a neuron is simple (see McCulloch and Pitts 1943). However, the description and manipulation of the huge networks in use today require deep mathematical knowledge. Through complex theoretical analyses and studies of the behaviour of these systems of simulated neurons, researchers have managed to establish the main properties of neural networks, which can be summarized as the ability to automatically classify data. That is to say, from statistical regularities detected in huge amounts of data, which are known as *patterns*, these types of system automatically create categories, in a process called *training*. The training process involves presenting the network with a series of examples for each category and adjusting the network after each example until it indicates the probability that a new item belongs to a specific category. For example, a DNN could be trained to recognize whether a sentence given as input is a positive financial forecast or a negative one. Later, when new data are presented, the network calculates the probability that this novel input belongs to one of the categories created during training (see Figure 9.1).

An Introduction to Narrative Generators. Rafael Pérez y Pérez and Mike Sharples, Oxford University Press.
© Rafael Pérez y Pérez and Mike Sharples (2023). DOI: 10.1093/oso/9780198876601.003.0009

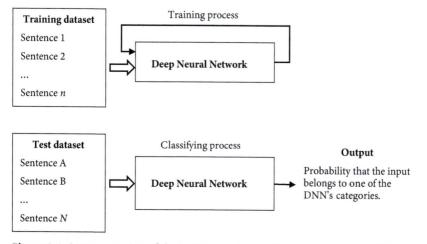

Figure 9.1 Representation of the training and classifying processes in a DNN

This chapter explains the working of a single neuron to introduce essential concepts. Then, in Chapter 10, we present a general description of DNNs and discuss how they are used for narrative generation. If you allow us to use a metaphor, first we will zoom in thousands of times with an imaginary microscope to study the behaviour of a single neuron; then we will zoom out to study the behaviour of an enormous network of these neurons that generates texts. In this way, by the end of Chapters 9 and 10, we will have the necessary elements to make an informed reflection on the scope and limitations of this approach to narrative generation.

9.2 Describing an artificial neuron

A neuron might have two or more inputs known as the *set of inputs*. Some of them might have a value of one, that is, they are active, and some might have a value of zero, that is, they are inactive. Each input is associated with a parameter called a *weight*, represented by the letter w, which has a profound effect on the firing conditions of the neuron (see Figure 9.2). The weight acts as an amplifier or inhibitor of the input pulse. The value of the sum of all active inputs multiplied by their respective weights is known as the *total input*, represented by the character sigma (Σ) in Figure 9.2, so Σ rises or falls as the input weights change. When Σ reaches a given threshold the neuron fires and the output is set to 1; otherwise, the output is set to 0.

Thus, the neuron behaves as follows:

If the value of the total input is greater than or equal to the value of the threshold, then the output of the neuron is set to 1; otherwise, the output is set to 0.

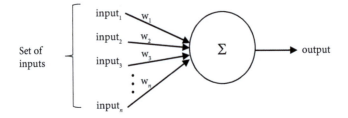

Total input = input$_1$*W$_1$ + input$_2$*W$_2$ + input$_3$*W$_3$ + ... input$_n$*W$_n$

Figure 9.2 Representation of an artificial neuron

This rule is known as the *activation function* and indicates the value of the output given a particular total input.

Neurons are employed as classifiers. A single neuron can only classify linearly separable data (if two sets of data points are represented in a two-dimensional space, and a line that completely separates both sets can be drawn, then those data sets are linearly separable). Let us illustrate this property. Imagine a dataset that can be divided in two types, known as A and B (e.g. the positive or negative financial forecast mentioned earlier). A neuron can be trained to correctly categorize the inputs from this dataset, known as the training dataset. If a given input belongs to type A (a positive financial forecast), the neuron's output must be 1; otherwise, it must be equal to 0. The goal of the training process is to make sure that the neuron behaves as expected. Thus, the main phases to train a neuron are as follows:

   i. Feed the neuron with an input from the training dataset.
   ii. If the input belongs to type A and the output is 0 (rather than the expected value 1), then increment the weights of all active inputs; if the input belongs to type B and the output is 1 (rather than the expected value 0), then decrement the weights of all active inputs; otherwise, analyse the next input.

These steps are repeated a predefined number of times. In this way, the training process focuses on modifying the weights to reduce the errors found.
   Thus:

- The learning process relies on knowing in advance which elements in the training dataset belong to type A and which elements belong to type B. This information is essential for training the neuron.
- Inputs must be represented as numerical values to determine whether the neuron fires.
- The behaviour of the neuron is adjusted by changing the weights of the active inputs.

Once the neuron has been trained, the system is evaluated employing a test dataset. As in the case of the training dataset, information about which elements in the test dataset belong to type A and which elements belong to type B is known in advance. The evaluation indicates how well the neuron is performing as a classifier.

9.3 How an artificial neuron works: classifying two types of sentences

In this section we develop an exercise to illustrate how a single artificial neuron works and to introduce some relevant concepts. The training dataset is composed of six sentences; the first three describe situations related to football/soccer; the last three describe situations related to family and park experiences:

Training dataset

1) The soccer stadium is beautiful.
2) My favourite soccer team plays in that stadium.
3) That soccer team is the best.
4) The park is beautiful.
5) My son plays in that beautiful park.
6) My son plays the guitar.

The goal is to train an artificial neuron to correctly classify the two types of inputs; those sentences that belong to the category family must fire the neuron while those texts that belong to the category soccer must produce 0 as an output. First, we train the neuron and then we check the neuron's capacity to correctly classify sentences.

9.3.1 Before starting the training process

The first step is to identify the vocabulary in the dataset. In this case, we have fifteen different words. As Figure 9.3 shows, each of these words is considered as one input and each has associated its own weight. The ordering of the words from one to fifteen is not important; the words are numbered here so we can refer to them later.

Next, we represent sentences as binary sequences. During the training process, the system will be given a series of sentences as input from the training dataset. For each sentence, the words that compose that sentence are set to 1 in the input set (i.e. they are active); the rest are set to 0 (i.e. they are inactive). In this way, only those words that are part of the sentence are considered when the algorithm adjusts the weights. Finally, we select at random the initial values for all the weights and the threshold. For the sake of clarity, for this exercise we set each weight to a value of 1, and the threshold to a value of 10.

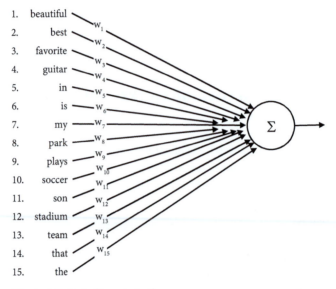

Figure 9.3 Set of inputs to the neuron

9.3.2 Training the neuron

To train the neuron we repeat several times a four-step algorithm. A variable called *time* is used to identify the number of cycles performed. Before the training process starts, the variable time is initialized to 0. The training works as follows:

1. The system selects the next sentence in the training dataset.
2. The system calculates the total input (it adds the current weight of each word in the sentence).
3. If this sentence belongs to the category 'family' *and* the output is 0 (i.e. the value of the total input is lower than the threshold of 10), then the system increments by 1 the weights of the words of the sentence; if this sentence belongs to the category 'soccer' *and* the output is 1 (i.e. the value of the total input is greater or equal to the threshold of 10), then the system decreases by 1 the weights of the words of the sentence; otherwise, the system skips step 3.
4. The value of time is increased by 1. Go back to step 1 until each sentence in the dataset has been processed.

In each cycle, the system processes one sentence. When all sentences in the dataset have been processed the system reaches what is known as an *epoch*. The designer of the systems decides the number of epochs to be executed during the training process. In this exercise we perform three.

9.3.3 Epoch 1

Table 9.1 illustrates how the value of the weights are modified during the training process for epoch 1. The first column contains all the words in the vocabulary; that is, it is the input set. The second column, labelled '$w_{t=0}$', shows the initial weights for all inputs; in this case all are set to 1. The subindex '$t=0$' should be read 'time equal to zero' and indicates that the training process has not started yet. As explained earlier, each time the system processes one sentence the variable time is increased by 1. Thus, because the training dataset comprises six sentences, when the variable time reaches the value of 6 the system completes the first epoch; when the variable time reaches the value of 12 the system completes the second epoch; and so on.

The following explains how each sentence is processed during epoch 1.

9.3.3.1 Processing the first sentence

Step 1. The first sentence in the dataset is 'The soccer stadium is beautiful'. The words 'beautiful' (input 1), 'is' (input 6), 'soccer' (input 10), 'stadium' (input 12), and 'the' (input 15) are active (note that the ordering of the words in the sentence is ignored). The rest of the words are inactive and therefore they are not considered during this cycle. The third column in Table 9.1 labelled $S1_{t=0}$ (S1 stands for sentence 1) shows the weights of all words that compose sentence 1 at time = 0; that is, these are the active inputs.

Step 2. The value of the total input is equal to 5 as shown on the bottom of the column $S1_{t=0}$. So the output of the neuron is equal to 0 (remember that the threshold is 10).

Step 3. Because the first sentence belongs to the category soccer and the output is 0, then the system goes to the next step. In other words, the system does not modify the weights because the neuron's output shows the expected value.

Step 4. The variable 'time' is set to 1 and the system goes back to step 1 to process the next sentence.

The fourth column in Table 9.1, labelled '$w_{t=1}$', shows the weights for all inputs at time = 1; in this case, because the weights are not modified, all have a value of 1.

9.3.3.2 Processing the second sentence

The second sentence in the dataset is 'My favourite soccer team plays in that stadium'. The system performs again the four steps. As shown in Table 9.1's column $S2_{t=1}$, the words 'favourite' (input 3), 'in' (input 5), 'my' (input 7), 'plays' (input 9), 'soccer' (input 10), 'stadium' (input 12), 'team' (input 13), and 'that' (input 14) are active. The value of the total input is equal to 8. Because sentence 2 belongs to the category soccer and the output of the neuron is equal to 0 the system does not modify the weights. The variable time is set to 2 and the system goes back to step 1 to process the next

Table 9.1 Inputs' weights during the training process for epoch 1

Weights

Set of inputs	$w_{t=0}$	$S1_{t=0}$	$w_{t=1}$	$S2_{t=1}$	$w_{t=2}$	$S3_{t=2}$	$w_{t=3}$	$S4_{t=3}$	$w_{t=4}$	$S5_{t=4}$	$w_{t=5}$	$S6_{t=5}$	$w_{t=6}$
1. beautiful	1	1	1		1		1	1	2	2	3		3
2. best	1		1	1	1	1	1		1		1		1
3. favourite	1		1	1	1		1		1		1		1
4. guitar	1		1	1	1		1		1	1	1	1	2
5. in	1		1	1	1		1	1	1		2		2
6. is	1	1	1		1	1	1	1	2	2	2		2
7. my	1		1	1	1		1	1	1	1	2	2	3
8. park	1		1	1	1		1	2	2	2	3	3	3
9. plays	1		1	1	1	1	1	1	1	1	2	2	3
10. soccer	1	1	1	1	1		1	1	1		2	2	1
11. son	1		1	1	1		1	1	1		2	2	3
12. stadium	1	1	1	1	1		1	1	1		1	1	1
13. team	1		1	1	1	1	1	1	1	2	2		2
14. that	1	1	1	1	1	1	1	2	1	2	2	2	3
15. the	1	1	1	1	1	1	1	2	2	2	2	2	3
Total input	**5**	**8**	**6**	**4**	**9**	**9**							

sentence. The column '$w_{t=2}$' in Table 9.1 shows the weights for all inputs at time = 2; again, because the weights are not modified, all are equal to 1.

9.3.3.3 Processing the third sentence

The third sentence in the dataset is 'That soccer team is the best'. The same process is repeated. Table 9.1's column $S3_{t=2}$ shows the weights of all active words. Because sentence 3 belongs to the category soccer and the output is equal to 0, the weights are not modified. The variable time is set to 3. The column '$w_{t=3}$' in Table 9.1 shows the weights for all inputs at time = 3.

   At this point we have processed all the sentences belonging to the soccer category. So far, our neuron is able to correctly classify them (i.e. for each 'soccer' sentence given as input, the neuron outputs 0). Next, the system processes the texts that belong to the category 'family'.

9.3.3.4 Processing the fourth sentence

The fourth sentence in the training dataset is 'The park is beautiful'. As shown in Table 9.1's column $S4_{t=3}$, the words 'beautiful' (input 1), 'is' (input 6), 'park' (input 8), and 'the' (input 15) are active. The value of the total input is 4 and therefore the output is 0. However, because this sentence belongs to the category 'family', the output should be equal to 1. To reduce the error, the system adds 1 to the weights of the words comprising the fourth sentence. In this way, hopefully the next time that the system processes this text it will be able to make the correct classification. The variable time is set to 4 and the system goes back to step 1 to process the next item in the training dataset. The column '$w_{t=4}$' in Table 9.1 shows the weights for all inputs at time = 4. This time, the weights of 'beautiful', 'is', 'park', and 'the' are equal to 2.

9.3.3.5 Processing the fifth sentence

The fifth sentence is 'My son plays in that beautiful park'. Table 9.1's column $S5_{t=4}$ shows the weights of words that compose the sentence. The total input is equal to 9, so the output is 0. Because this sentence belongs to the category 'family', the output should be equal to 1. To reduce the error the system adds 1 to the weights of the words composing the fifth sentence. The variable time is set to 5 and the system goes back to step 1 to process the next sentence. The column '$w_{t=5}$' in Table 9.1 shows the weights for all inputs at time = 5. The weights of 'beautiful' and 'park' are equal to 3, while the weights of 'in', 'my', 'plays', 'son', and 'that' are equal to 2.

9.3.3.6 Processing the sixth sentence

The sixth sentence is 'My son plays the guitar'. As shown at the bottom of Table 9.1 column $S6_{t=5}$, the total input is equal to 9, so the output is 0. The system adds 1 to the weights of the words comprising the sentence. The variable time is set to 5 and the system goes back to step 1 to process the next sentence. The column '$w_{t=6}$' in Table 9.1 shows the weights for all inputs at time = 6. The weight of 'guitar' is 2, the weights of 'my', 'plays', 'son', and 'the' are equal to 3.

   At this point the system ends the first epoch.

Table 9.2 Value of the total input for each sentence during training

	Sentences describing soccer topics			Sentences describing family experiences		
	Sentence 1	Sentence 2	Sentence 3	Sentence 4	Sentence 5	Sentence 6
Epoch 1	5	8	6	4	9	9
Epoch 2	10	12	4	8	16	12
Epoch 3	6	4	6	12	16	12

Note: Entries with a grey background indicate incorrect outputs.

9.3.4 Epochs 2 and 3

The four steps described in Section 9.3.3 are repeated for each sentence until a pre-determined number of epochs is reached. For this exercise we have established three epochs. Table 9.2 shows the value of the total input for each sentence during training. Entries with a grey background reveal incorrect outputs. This information makes it possible to perform a general analysis of the behaviour of the neuron. During epoch 1 those sentences that belong to the category 'soccer' are correctly classified while those texts that belong to the category 'family' are mistakenly catalogued. This makes sense because the initial values of all weights are 1, so at the beginning of the process it is difficult to reach a total input of 10. During epoch 2, the adjustments of the weights result in sentences 1, 2, and 4 being mistakenly catalogued, although sentences 5 and 6 are now correctly classified. Finally, during epoch 3 the system finds the precise value for all weights so all sentences are correctly classified. If the training process were to continue, the weights would not be modified further.

9.3.5 Analysing the final weights

Table 9.3 shows the weights for each epoch. The analysis of this information allows some conclusions to be drawn:

(a) The training algorithm allows the neuron to identify key features for each category. Words like 'soccer', 'stadium', and 'team' are distinctive elements of the category 'soccer'; that is why these words ended with low, or even negative, values of weights, as shown in Table 9.3. Assigning low values is how the system makes sure that the neuron does not fire when key words of the category 'soccer' are present in the input sentence. On the other hand, 'son' and 'park' are distinctive elements of the category 'family'; that is why these words ended with the two highest values of weights. Assigning these high values is how the system makes sure that the neuron fires when key words of the category 'family' are present in the input sentence.

Table 9.3 Inputs' weights for each epoch

Words	$w_{t=0}$	Epoch 1 $w_{t=6}$	Epoch 2 $w_{t=12}$	Epoch 3 $w_{t=18}$
1. beautiful	1	3	3	3
2. best	1	1	1	1
3. favourite	1	1	0	0
4. guitar	1	2	2	2
5. in	1	2	1	1
6. is	1	2	2	2
7. my	1	3	2	2
8. park	1	3	4	4
9. plays	1	3	2	2
10. soccer	1	1	−1	−1
11. son	1	3	3	3
12. stadium	1	1	−1	−1
13. team	1	1	0	0
14. that	1	2	1	1
15. the	1	3	3	3

However, claiming that the neuron is able to identify key features of a category is a metaphorical expression in the sense that the system does not possess any kind of representation of the meaning of concepts like soccer or park; they are just inputs whose weighs are modified based on statistical regularities found in the training dataset.

(b) Several factors influence the value of the final weight for each input. For instance, the number of words in a sentence impacts the value of the total input. If we employ long sentences for the category 'family', then the process of reaching a high value for the total input would be faster. As a result, the weights of the words of those long sentences would be modified less frequently.

Although the word 'the' appears twice in the category 'soccer' and twice in the category 'family', it ends with a weight equal to 3, which is the second highest value in Table 9.3. Because this word is present in sentences from both categories, one would expect its final weight to be a medium value of 1 or 2. However, this is not the way the algorithm works. The algorithm focuses on increasing or decreasing the weight of the active inputs. Other information, like the relation between words and categories, is not considered.

9.4 Testing the neuron

The next step is to test how the system works. In this case, the neuron is able to correctly classify sentences not included in its training dataset. For instance, 'My son plays soccer in that stadium' reaches a total input of 7 and therefore it is correctly classified as belonging to the category 'soccer', while the sentence 'The beautiful guitar is my favourite' reaches a total input of 12 and therefore it is correctly classified as belonging to the category 'family':

New sentence	My	son	plays	soccer	in	that	stadium	
Weights	2	3	2	−1	1	1	−1	Total input = 7

New sentence	The	beautiful	guitar	is	my	favourite.	
Weights	3	3	2	2	2	0	Total input = 12

The neuron is able to classify sentences that include words that are not part of the vocabulary. The simplest way to implement this feature is to develop a routine that eliminates from the input those unknown items. In this way, the sentence 'My oldest son plays soccer every morning in that stadium' is transformed into 'My son plays soccer in that stadium'; then it is classified as belonging to the category 'soccer'.

However, because this exercise involves a very small training dataset, the neuron can easily misclassify some inputs. For instance, 'My son plays soccer in the best team' has a total input of 11 and as a result, it is mistakenly considered as belonging to the category 'family'.

New sentence	My	son	plays	soccer	in	the	best	team	
Weights	2	3	2	−1	1	3	1	0	Total input = 11

This is a particularly complicated sentence because 'son' and 'soccer' are two key words that belong to different categories. Thus, to figure out the right classification is problematic.

The scope of the neuron in this example is very limited. There are endless expressions describing soccer or family situations that are not covered by this system. The extent of the vocabulary (i.e. the set of inputs) and the number of statistical regularities that can be found during training depend on the size and quality of the dataset and in the number of epochs performed.

The neuron has no means of evaluating the coherence of the input. The sequence of words 'That my in' has a total input equal to 4 and therefore it is classified as belonging to the category 'soccer'. The sequence 'That my is in best beautiful' has a total input equal to 10 and therefore it is classified as belonging to the category 'family'. So, the correct operation of the neuron relies on the quality of its inputs.

In summary:

- The term *training* refers to the process of mechanically modifying the weights of the inputs.

- The term *learned* refers to the situation where the weights have reached a value such that the system is able to correctly classify the majority of the inputs in the training dataset. That is, the system learns to generate a particular output (or pattern of activation) whenever another particular input pattern occurs (Rumelhart and McClelland 1986, 54).
- The training process allows the neuron to detect statistical regularities in the dataset. That is, the system is able to detect key features for the categories represented in the inputs.
- The quality and size of the dataset is essential for the correct working of the neuron.
- The system has no means of evaluating its inputs.

Working with a neuron consists of three phases. In the training phase, a training dataset (where the output for each entry is known) is used to modify the weights and to produce the desired output. In the testing phase, the system is assessed by employing a test dataset (where the output for each entry is known). Finally, in the performance phase, the system is used to classify novel inputs where the output is not known in advance.

9.5 Final remarks about neurons

As mentioned earlier, individual neurons can only classify linearly separable data. This is a severe shortcoming because numerous situations in the real world are not linearly separable problems (e.g. identifying computer malware). However, this drawback can be overcome by organizing neurons in layers and by employing non-linear activation functions to trigger neurons. As shown in Figure 9.4, in this type of system, inputs are connected to neurons in the first layer, which in turn are connected as inputs to the second layer, and so on; the designer decides the number of neurons and layers in the system.

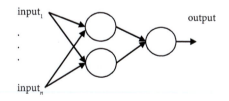

Figure 9.4 A two-layer system

When the outputs of the neurons in the first layer are used as inputs for the neurons in the second layer, the system can process more abstract data. For instance, the output of the neuron in the previous exercise indicates whether a sentence describes a soccer or a family situation. This output can work as an input to a second layer to produce more general classifications. Neurons organized in multiple layers are known as neural networks. We will study them in Chapter 10.

10
Inspired by the brain

Deep neural networks

10.1 Introduction to deep neural networks

In recent years, technological developments have produced increasingly powerful computers, resulting in the ubiquitous interconnectivity that we experience, the digital irruption in practically all aspects of our lives that, among other things, have led to the digitization of huge amounts of data. The desire of big technology companies to find ways to commercially exploit all this data has given a boost to the research and development of deep neural networks (DNNs). Research efforts in this area have mainly focused on: (1) developing mechanisms that control how and when a neuron fires; and (2) designing various ways for these neurons to organize and communicate with each other. It is expected that through manipulation of these features, the models will perform complex tasks, such as accurately translating text from one language to another. In contrast with the computer programs we have seen in previous chapters, DNNs employ huge amounts of interconnected, relatively simple information-processing units, which evoke the functioning and organization of the brain, to represent in a distributed way information that the system requires to function.

It is beyond this book's objectives to explain the nature of distributed parallel processing, the characteristics of the mathematical models used in DNNs, or the various ways that these networks can be organized. Thus, we describe the types of processes that DNNs might execute but we do not give details on how the networks carry them out (for details see e.g. Aggarwal 2018, Goodfellow et al. 2016).

10.2 General description of deep neural networks

In the 1980s and 1990s, a typical neural network comprised hundreds of elements organized in layers. Nowadays, deep neural networks interconnect millions of those artificial neurons. Although these programs are highly complex, the concepts introduced in Chapter 9 are useful for describing some of the core features of a DNN.

DNNs are mathematical models. Thus, linear algebra, in particular vectors, is useful for describing some of their properties. A vector can be described in simple words as a pattern of numbers. If your best friend is 25 years old, is 170 cm tall, and weighs

An Introduction to Narrative Generators. Rafael Pérez y Pérez and Mike Sharples, Oxford University Press.
© Rafael Pérez y Pérez and Mike Sharples (2023). DOI: 10.1093/oso/9780198876601.003.0010

60 kg, you can represent these features as a vector as follows: [25, 170, 60]. A vector can represent as many properties as we want. In a DNN, several components such as inputs and weights are represented as vectors.

Typically, as shown in Figure 10.1, a network is made up of a group of input neurons, a group of intermediate neurons organized in layers known as hidden layers, and a group of output neurons. All neurons in the net have associated weights. The stimuli that activate the network are received through the input neurons; the neurons in the intermediate layers propagate the input stimuli throughout the entire system; the output neurons provide the result that the network produces. The neurons in the intermediate layers are known as 'hidden' because they do not connect directly to input or output. An intermediate neuron can receive hundreds of thousands of impulses as inputs and, in turn, its output might be connected to a similar number of neighbours. The rectangles in Figure 10.1 represent that the hidden layers can have different architectures and goals, depending on the purpose of the system.

Rather than only working with discrete values, like the neuron in the example from Chapter 9, DNNs work with continuous values. Thus:

- DNNs employ nonlinear activation functions to calculate the value of the output for each neuron. Typically, these values are real numbers in the range between −1 and 1, or in the range between 0 and 1.
- This feature provides a great flexibility for adjusting the weights during training.
- When the dataset does not consist of numeric data, it is necessary to transform it into numbers; typically, words are represented as vectors.
- Because the outputs of DNNs might have values between 0 and 1, those outputs can represent probabilistic values.

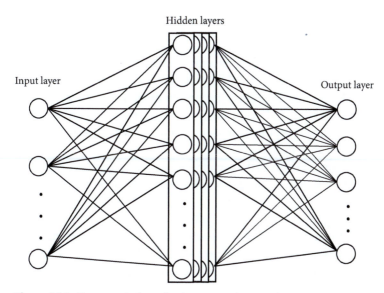

Figure 10.1 Representation of a deep neural network

In a parallel distributed system, what causes a network to behave in one way or another is the combination of the input stimuli, the architecture used to build it, and the values of the weights associated with each neuron. Because during training neither the architecture nor the dataset is modified, the only way to alter the behaviour of the network is by modifying the weights. Thus, training basically consists of finding the right combination of weights to produce the desired output.

The first step in building a deep neural network is to design its architecture, that is, to define how many neurons the system will have, how they will be organized, and how they will communicate to each other. Next, the DNN is trained. The goal of training is to get the network to establish a relationship between a given input and a desired output. That is, when the network receives a specific input it must generate an expected output. If the network does not produce this output, its internal parameters—which control an important part of the system's operation—are modified to reduce the error; that is, they are adjusted to reduce the difference between what the system is expected to produce and what it is actually producing. This process is repeated until, hopefully, the error disappears. This process is repeated a predetermined number of epochs. It is common that the output for each entry in the training dataset is known in advance, for example, because humans have labelled each entry in the dataset with the correct output. The process of training a DNN with this type of dataset is known as supervised learning. As a result of this training process, the network creates categories like those we explained in Chapter 9. Thus, when a new input is given to the DNN, the system outputs the probability that such an input belongs to one of the registered categories.

In Chapter 8 we explained that generalization consists of identifying from a set of input patterns those features that are significant. In this way, new patterns can be associated with others that share the same distinguishing characteristics. DNNs have strong capacities for generalization; the training process modifies the weights in such a way that significant features are clearly identified.

Thus:

- The term *training* refers to the process of mechanically modifying the weights inside the network. As one would expect, the training process in a DNN is much more elaborate than for a single neuron. In modern systems, researchers have developed sophisticated mathematical models for updating the weights. However, the essential purpose is the same: to modify the weights in order to reduce the error between the input and the desired output.
- The term *learned* refers to the situation where the weights have reached a value such that the system is able to correctly classify the majority of the inputs in the training dataset.
- The training process allows the DNN to detect statistical regularities in the dataset. The use of multiple layers allows it to detect elaborated patterns at different levels of abstraction that represent key features for the categories represented in the inputs (e.g., to learn patterns of grammar from a training dataset of texts).

- The quality and size of the dataset are essential for the correct working of the DNN.
- DNNs have no means of evaluating their inputs for accuracy or coherence.
- The output of a classifier DNN is the probability that a given input belongs to one of the defined categories.

10.3 Deep neural networks for text generation

Besides performing classification tasks, such as detecting whether a passage was written by Jane Austen or Charles Dickens, researchers have succeeded in creating DNNs that employ probabilities to generate new texts. The following is the core ideas behind these narrative generators. The training dataset consists of a corpus of human-written texts. Because the complete writings are available then the system can determine which word follows any sequence of words in the training dataset. That is, rather than calculating the probability that a given input belongs to a category, the system calculates the probability that a given input is followed by each word in the corpus (this probability is similar to what we studied in Chapter 8; e.g. see Tables 8.1 and 8.2). In this way, instead of having a discriminative model we have a generative model where the output is the next word in the sequence.

Because in text generators the output for a given sequence of words is part of the input, no human labelling is required; the process of training a DNN with this type of dataset is known as self-supervised learning. A basic example is the text completion software on a mobile phone or word processor. Given a word or phrase, such as 'Thank you', it can suggest a continuation, such as 'for the'. The text generators we discuss here are more highly trained and sophisticated descendants of that approach.

Text generators present particular challenges for researchers. For instance:

(a) *Building the dataset.* The construction of the dataset is a task that involves not only searching for and obtaining the texts, but also cleaning them up. For various reasons, many of the writings available on the Internet include what is known as 'trash' or 'noise'. Typical cases involve finding non-text characters like 'ᄇ', punctuation marks in incorrect positions, incomplete phrases or sentences, and words with missing or altered letters. It also happens that texts extracted from social networks include misspellings or abbreviations not acceptable in formal language. Network designers must also ask questions like, should the dataset be converted into lowercase? This is useful because computers are case-sensitive and therefore for them the word 'Time' is not the same as 'time'. However, capital letters serve a relevant function in language, such as when they are used to indicate proper names. So any determination can lead to unwanted consequences. Like these, there are many other choices that must be made. Many of these tasks to clean a dataset can be automated, but the procedures to do that need to be carefully designed. So building a powerful dataset is a laborious task.

(b) *Representation of words.* In the example in Section 9.3, each word is represented as a single neuron that is active or inactive. A DNN requires a richer mathematical representation, like vectors, which allows it to perform more elaborate processes. Thus, it is necessary to develop a mechanism that can transform words into vectors.[1]

(c) *The system must be able to consider previous data.* Most storytellers based on DNNs produce paragraphs that are grammatically accurate; however, when the output text begins to grow, these systems have problems maintaining coherence in the text (Martin et al. 2017). That is, the semantic and syntactic relationships between what happened, what is happening, and what is going to happen in the tale need to be consistent. Researchers in the area call these relationships *dependencies.* However, DNNs have trouble representing such dependencies because a good story needs a coherent overall structure, but language models generate text word by word; they do not operate with high-level plans or goals (Fan et al. 2018). As a result, these programs may not produce well-connected, or even related, sequences of paragraphs. Let us elaborate this idea. As we pointed out earlier in this book, texts have a strong dependency between what has happened and what happens next. If in the expression 'Scientists were capable of finding a cure to this new disease in a record time', the word 'capable' is changed for 'incapable', the whole meaning of the sentence and, therefore, the way it should continue, changes. A narrative generator that considers only the last words of the sentence during the generation process will miss this vital information. Here is another example. 'The Beatles composed great songs. This band made an impression on me'. In the second sentence, the word 'band' refers to 'The Beatles'. This type of contextual information is relevant to coherently progress the text; in this case, because we know that the word 'band' refers to a musical group rather to a squad of criminals, we can find an appropriate description to continue the narrative.

10.3.1 Layers in a DNN

To deal with some of these challenges, researchers have designed diverse DNN architectures that are organized in layers (see Figure 10.1). Layers can be pictured as modules in an assembly line; the output of one layer is the input of the next layer. Each of them has different features and has been designed to achieved different goals. The following describes the general features of three commonly used layers for narrative generation: the embedding, the long–short term memory, and the dense.

(a) *The embedding layer.* The embedding layer transforms each word in an input sequence into a vector. The length of the vector is a decision of the designer of the system. Besides the necessity of representing inputs in mathematical

---

[1] Some DNNs work with letters or syllables rather than words.

terms, there is another important reason to represent words in this way. Researchers have developed mechanisms that, when the meaning of two words is related, their representation as vectors is alike. In this way, the similarity between vectors shows the semantic and syntactic similarities between words (Mikolov et al. 2013). So the system can identify that the word 'taco' is related to the word 'quesadilla'. DNNs for narrative generation exploit this type of features. There are tools, like the project Global Vectors for Word Representation,[2] which can be downloaded and used to obtain representative vectors for an enormous set of words.

(b) *The long–short term memory (LSTM) layer.* Systems based on probabilistic distributions, like DNNs, find it hard to identify long-distance dependencies, particularly when there is a large gap between the words in a phrase that are connected. Researchers have developed different architectures to attempt to solve this problem with different degrees of success. A popular one is the long–short term memory (Hochreiter & Schmidhuber 1997), which has been developed to work with sequential data like texts. The main goal of a LSTM is to address the problem of long-term dependencies by providing memory to the system. Thus, although during the training and classification the system processes one word after another, it is able to remember and employ information about previous words in a given sequence.

(c) *The dense layer.* One of the most common layers is known as the dense or fully connected layer, where all neurons in the layer are connected to all the neurons in the previous layer. It focuses on the classification part of the process. There are diverse activation functions that can be used with dense layers. When a dense layer is used as an output, it is common to employ the softmax activation function. This function makes the total output of the layer equal to 1. So it is useful for calculating probabilistic distributions.

In this way, what makes a DNN a powerful tool is the combination of the following features:

- Capacity to represent millions of words;
- Capacity to recognize similar words during training and during text generation;
- Capacity for automatic pattern recognition;
- Capacity to generalize; and
- Capacity to use multiple layers, each with its own architecture and purpose, that work together in an analogous way to an assembling line, to generate a narrative.

[2] https://nlp.stanford.edu/projects/glove/

10.3.2 Designing a narrative generator

Thus, to develop a narrative generator, it is necessary:

- to define the architecture of the DNN;
- to build a dataset;
- to train a DNN for text generation; and
- to generate new narratives.

For this exercise, we employ a simple DNN to propose the next word in a text it is given as input. It includes the three layers described earlier: one embedding layer, one LSTM, and one dense layer that employs a softmax activation function.

The training process is similar to what we have described in previous sections. The inputs are sequences of words from the training dataset represented as vectors. Each neuron in the output represents one of the words or punctuation marks in the training dataset. The system calculates the probabilistic distribution for each sequence of word used as input. For this example, we are going to assume that the training set contains a total of 4,500 different words and punctuation marks, so the output has the same number of neurons (a real DNN will have a much bigger training set).

Figure 10.2 represents a DNN already trained to produce sequences of words. So the generation process from a trained network would work as follows:

1. The user chooses as input a sequence of words, for example, 'Today little Hans'.
2. The embedding layer transforms this sequence of words into vectors that are used to feed the next layer. For this example, each word is represented as a vector of size 512 (the size of the vector is decided by the designer). Thus, the system employs one neuron to represent one element in each vector; that is, 512 neurons represent one word. If we design a system that can read five words, then the input requires 2,560 neurons.
3. Data percolates through the LSTM; each layer outputs a vector that includes information about the current and previous words, and which is used to feed the next layer.
4. The output of the DNN is a dense layer that employs a softmax activation function that consists of 4,500 neurons, each representing one of the words or punctuation marks found in the training set. Each of the output's units can have a value between 0 and 1, which represents the probability that that word will follow the input sequence. So the output of the network is the probabilistic distribution of all the words and signs that can follow the input sequence. In this case, 'Today little Hans' has a 39% probability to be followed by ",, a 12% probability to be followed by 'was', and so on. This is the kind of probability distribution discussed in Chapter 8.
5. One of the options is selected. For this exercise, we choose at random one word or punctuation mark from those with the ten highest probability values, for example, 'was'. This item is placed at the end of the input sequence, and the process is repeated until the generated output text reaches a predefined length.

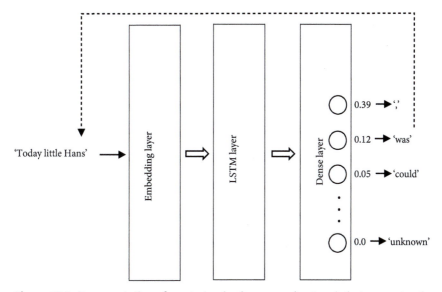

Figure 10.2 Representation of pre-trained a deep neural network that generates the next word in a sentence

This example shows just one of multiple architectures that might be used for text generation. A promising alternative is known as Transformer (Vaswani et al. 2017). This was initially used for machine translation and automated question answering, but has since been applied to text summarization and story generation. Like LSTMs, transformers are designed to handle sequential data such as sentences, but instead of processing the items in order, one word after another, it models relationships between all the words in a sentence. Let us say we have a program to generate book recommendations. How it completes the sentence 'This superb memoir is a revealing and stimulating view . . .' depends on the meaning of 'view', which is influenced by the word 'memoir' occurring six words earlier. Ending the sentence with 'view of a troubled childhood' is more appropriate than, say, 'view of a changing landscape'. For other types of DNN architecture, the further that words are separated, the harder it is to determine how they relate. By contrast, transformers capture relationships between all the words in long passages at once, regardless of their position. The basic mechanism of a Transformer is called 'self-attention' (Bahdanau et al. 2014). It lets all the input words interact with each other and finds out which ones the system should pay most attention to. In this case, 'memoir' would get a high attention score when working out how to follow the word 'view'. We can test how this works in practice. If we give a Transformer program the input text 'This superb *memoir* is a revealing and stimulating view of', the program continues with 'how one of the great minds of our time experienced the world'. If instead we start with 'This superb *image* is a revealing and stimulating view of', the program adds 'the iconic building. The subtle lighting adds to the atmosphere of this excellent photograph.'[3]

---

[3] The Transformer program is GPT-3, which we discuss later in Section 10.4.4.

10.4 Some illustrative real systems

Now it is time to study some real systems. Researchers in the area give special attention to solving the coherence and dependency problem. In this section, we review some of the solutions that experts in the field are exploring. When available, we present examples of the produced texts. This should be considered as an illustrative rather than an exhaustive list of narrative generators.

10.4.1 A thematic reference

Fan et al. (2018) have developed a system that generates stories around a premise. The system first produces a sentence, called a prompt, which we can describe as a thematic reference that helps the neural network to focus on a topic. Next, when generating a narrative, the prompt acts as a guide that conditions the selection of words to progress the text. The authors obtained from an online forum more than 300,000 human-written stories paired with their respective prompts; each prompt might be linked to more than one story. They use these data to train their system. Thus, by learning the relation between the prompt and the story and employing a self-attention mechanism, this neural network is able to produce a narrative that progresses around the given prompt. At each step the program generates the probability of each word in the vocabulary being the likely next word. Here is an example published in Fan et al. (2018) that illustrates the kind of output that this system generates.

> Prompt: Aliens start abducting humans
>
> 'It has been two weeks, and the last of my kind has gone. It is only a matter of time until there will be nothing left. I'm not sure what the hell is going on . . . I can't think. I can hear a distant scream. I think of a strange, alien sound. I try to speak, but am interrupted by something, something that feels like a drum, I ca not tell. I mean I'm just a little bit older than an average human. But I can, and I can feel the vibrations. I hear the sound of a ship approaching. The ground quakes at the force of the impact, and a loud explosion shatters the silence.'

10.4.2 Employing narrative structures

In Chapter 8 we studied how employing sequences of words (idioms or even whole sentences) produces better results when progressing a text than using only individual words. However, their use carries several complications. The main one is that writers attempt to avoid repeating sentences. Also, because DNNs require huge datasets in order to capture the recurrence of elements, it can be expensive to train a network that generates the probabilistic distribution of sentences. The solution that Martin et al. (2017) propose is to convert sentences into a more abstract representation, which they call *events*. That is, the objective is to transform diverse expressions into the same event and in this way train a network capable of establishing probabilistic relationships between events. The authors developed a system that, employing natural language techniques, transforms a written story into event-structures. Next, they train a neural network to calculate the probability distribution of the next event to occur. In this way, their system can generate new narratives. An event is represented as a structure that comprises four elements: subject, verb, object, and modifier. The object and the modifier are optional. In this way, the sentence 'Marcia went to the beach' is transformed into <marcia; go; beach;> (in this example there is no modifier). Complex sentences can be transformed into two or more events. For example, 'Joe and Tania went to the theatre' becomes <joe; go; theatre;> and <tania; go; theatre;>. The system can generate general representations of events by substituting more generic descriptions for specific ones; for example, the system might employ 'self-propelled vehicle' instead of 'car'. When the generation process ends, the authors employ a second neural network that attempts to transform the chain of events that represents the new story back into text that people can read. However, converting abstract sequences of events into a coherent written story has proved to be a very complex problem that the authors have not yet been able to solve. So, they do not provide an example of an output as a written tale. Nevertheless, this work illustrates a generative NN that works on event structures rather rather than surface text. The authors claim that since events can represent basic semantic information from the sentences of a corpus of stories, then their system is capable of learning the skeleton of a good story and can use such a template to generate novel stories.

10.4.3 Planning

Tambwekary et al. (2019) model story generation as a planning problem. Their main goal is for the system to develop a story that ends in an event where a previously chosen verb occurs. Thus, the program starts with initial and final story-states; then the system attempts to find the series of events that transforms the initial state into the goal state. For instance, given as an initial incident 'Victoria hated to walk in the park' and as a goal the verb 'marry', the system must find a sequence of events that ends in someone (maybe the same Victoria) getting married. The authors use the work of Martin et al. (2017) mentioned earlier as a starting point. That is, they begin with a neural network trained to generate sequences of events. Next, they continue training this neural network with a *reward function* whose objective is to predispose the system to choose verbs closer to the goal. Then the system is ready to generate narratives.

The reward function is based on the idea that there are actions in a narrative that are more likely to appear closer to the target action than others. For instance, situations involving the verb 'meet' probably are closer to the goal of admiring a character than those including the verb 'leave' (perhaps because it is easier to admire someone if you meet that person). With this kind of information, during the progression of a narrative, the neural network can choose from the available options the one that moves the tale closer towards the goal. To build this reward function, before running their neural network, the authors analyse the dataset to determine how far a given verb is from the desired action, namely, how many events separate a specific state from a given goal. They also take into consideration how frequently a particular verb appears before that target action. The results of this analysis determine how to reinforce the weights in the system in order to guide the generation of the plot to a given goal. Because this system is based on the representation of events developed by Martin et al. (2017), it also cannot produce written tales yet. Nevertheless, this system illustrates a NN for narrative generation based on basic problem-solving concepts.

10.4.4 Going massive

In June 2020, the OpenAI company released GPT-3, a Generative Pre-trained Transformer with 175 billion parameters (Brown et al. 2020). That was followed in January 2022 by InstructGPT, a version that accepts direct instructions from users. In November 2022, the company launched ChatGPT, a conversational model based on an advanced version of GPT-3, followed in March 2023 by GPT-4. These systems have had strong coverage in national newspapers, blogs, and social networks. As we explained earlier, transformers are designed to handle sequential data such as sentences, but instead of processing the items in order, one word after another, they model relationships between all the words in a long passage at once (for GPT-4, approximately 25,000 words).

The GPT systems exhibit two peculiarities that distinguish them from the programs we have reviewed so far. The first is its colossal size. These systems employ up to 1 trillion parameters (input weights for the artificial neurons). To train GPT, the company has used a dataset derived from multiple sources, for example, blogs, social networks, and digitized books collections, that practically reach trillions of words. Few organizations have the computing capacity necessary to process this amount of information. The second distinctive feature is the purpose of the programs. Unlike the other systems we have described, GPT has not been built solely for the purpose of generating narratives; this is one of the many tasks it can do. Its creators have evaluated the system in various activities such as predicting the best word to finish a sentence, choosing the best ending for a five-sentence story, answering questions, translating texts from one language to another, and determining the comprehension of a reading, among others. In this way, this project aims to contribute to understanding the possibilities of using the same and complex parallel distributed processing architecture, and the same dataset, to perform multiple tasks. Here we focus on narrative generation. We instructed GPT-3 to generate two stories: one about a Mexica goddess called Coatlicue and one about the foundation of Tenochtitlan (which today is México City). These are the results:

Instruction to GPT-3: Write a new version of the story about the Coatlicue.
In Aztec mythology, Coatlicue is a mother goddess. The people of the Aztec tribes believed that she was responsible for their existence. One day, she became

pregnant and gave birth to her youngest son. She loved him very much and promised to take care of him, but he eventually died at an early age. When he did, Coatlicue became very sad and went into a deep depression. Her pain was so strong that it caused her to produce another child from her womb—a baby girl called Coyolxauhqui. Coatlicue reluctantly took care of this child and fed her with the milk from her breasts. This time, when Coyolxauhqui grew up, she became jealous of all the attention that Coatlicue was giving to her other children and tried to kill them using snakes as weapons. Despite this, Coatlicue still loved Coyolxauhqui and felt sorry for her when she saw what she had done—so much so that she even made the weapons turn into flowers.

Instruction to GPT-3: Write an alternative version of the foundation of Tenochtitlan.
It is said the Aztecs built Tenochtitlan on the site of an old settlement. One day, a group of Aztecs were out in the fields hunting and had just killed a deer. They began to make their way back to Tenochtitlan through the forest when they stumbled across a nest of snakes. The Aztecs killed the snakes and then took their skins to Tenochtitlan. When they arrived at Tenochtitlan, they told the people of their adventure and the people rejoiced with them as they had been in need of new clothes for years. The people dressed up in the skin of the snakes that had been killed and from then on, it was believed that snakes lived inside every Aztec.

The GPT systems illustrate the power of transformers in narrative generation.

Although large language models have an enormour influence on society, few researchers and developers have full access to these tools due to the huge costs involved in training them. Various initiatives that seek to alleviate this problem are emerging. An informative example is BigScience,[4] which developed the BigScience Language Open-science Open-access Multilingual (BLOOM) language model. BLOOM works in a similar way to GPT-3 but it has been trained on forty-six different languages.

This project is the result of the collaboration of over 1,000 researchers from more than 70 countries, and more than 250 institutions. Their goal is that researchers, students, educators, developers, and non-commercial entities, basically any person interested, can work with BLOOM. Similarly, some researchers are employing tools like Hugging Face,[5] an open repository, for sharing and exploring transformer systems. Even some private companies have developed transformer models that have been made available to the research community, for example, OPT (Zhang et al. 2022). Hopefully, all these efforts will contribute to developing a more equal and inclusive AI society.

---

[4] https://bigscience.huggingface.co
[5] https://huggingface.co/docs/transformers/index

10.4.5 Combining different approaches

Nye and his colleagues (2021) developed a system inspired by the idea that human cognition is a constant back-and-forth between an intuitive and associative system and a more deliberative and logical system (Evans 2003). Nye et al. call the former System 1, and the latter System 2. In this way, the function of System 1 is to produce material for the development of the narrative, while System 2 evaluates the coherence of said material and, if necessary, discards it. These ideas are very similar to those we will present in Chapter 11, where we introduce the cognitive account of creative writing known as engagement and reflection (Sharples 1999).

Nye and his colleagues use DNNs to implement System 1. They choose this approach because they find similarities between the erroneous 'gut' (or intuitive) answers that humans give when asked questions on reasoning tests, and the answers that GPT-3 produces to the same questions. The authors suggest that inconsistencies and lack of logic that sometimes appear in texts produced by the neural models of language might be the result of processes that represent the 'intuitive' generation of sentences. To implement System 2, the authors use a symbolic model based on rules. The main goal of System 2 is to create a world model that helps to verify whether the statements that System 1 produces are consistent with such a world. This model includes basic information such as where each character is located, what objects each one possesses, and what family relationship one character has with another. The model also includes a set of actions that can be performed by characters, such as 'go', 'drop', 'pick up', and some common-sense rules, such as a person cannot be in two different places at the same time, the same object cannot be in the possession of two different people, a character's grandfather cannot become his brother in the next action, a character cannot go to the same place where it is currently located. All this information is encoded by the authors as rules. The System 2 also checks that the preconditions to perform an action are fulfilled; for example, if a character is going to pick up an apple, both must be in the same place.

In general terms, the system works as follows:

- System 1 is a pre-trained language model which generates an initial short text. For example, 'Antonio went to the movies. Lucy went to the beach. Antonio had a soda.'
- Based on this text, System 2 creates a model of the world. In this example, this model describes the location of Antonio and Lucy, and who is holding the soda: Antonio.Position = cinema; Lucy.Position = beach; Soda.Holder = Antonio.
- System 1 generates a group of sentences, each of which is considered a candidate for continuing the text. For example: 'Lucy dropped the soda', 'Antonio returned home.'
- System 2 transforms the candidate sentences into a logical representation: drop (Lucy, soda); go (Antonio, home). This is necessary because System 2 is a rule-based system.

- System 2 verifies that the first candidate, represented as drop (Lucy, soda), is consistent with the current state of the world. If the sentence is coherent, it is added to the text in progress. If it does not meet the consistency requirement, the sentence is deleted and the process is repeated with the next candidate. In this example, Lucy can't throw away the soda because Antonio is the one who has it. Therefore, this candidate sentence is eliminated. The fact that Antonio returns home, represented as go (Antonio, home), does not break any coherence rule and is appended to the end of the text in progress. In this way, the system generates the following: 'Antonio went to the movies. Lucy went to the beach. Antonio had a soda. Antonio returned home.'

To evaluate their system, the authors asked eighty human judges to participate in a test. Participants were shown a prompt consisting of several sentences and were asked to choose which of two possible continuations made the most sense. One of them was generated by a language model, and the other was generated by the dual system described here. Each judge evaluated between twenty and twenty-six trials. The authors report that, in the vast majority of cases, the human judges selected the sentence generated by the authors' system as the most coherent.

This project illustrates how rule-based systems and DNNs might work together. Its main limitation is how basic its representation of the world is. Creating a rich model of the world that can be used in the variety of situations that language models generate is a complex task.

10.5 Final remarks about deep neural networks

Designers of DNNs struggle to explain the inner workings of their networks. Because of the huge number of elements interacting in this type of system, it is impossible to analyse in detail the rules generated during training that dictate the behaviour of the network. Some commonly reported limitations of narratives generated with neural networks are the gradual loss of coherence over long passages, the production of contradicting statements, the repetition of sequences, and the generation of off-topic sentences. This approach requires hundreds of thousands of examples to generalize well.

Storytellers based on DNNs have achieved what few other systems studied in this book have done: the use of words as vital elements to shape the output during the creation process. Let us elaborate this idea. The methodologies analysed in other chapters represent narratives as related structures that characterize events, characters, goals, and so on; once the narrative is built, these structures are 'translated' into a readable format using templates. In this way, the expressions chosen to describe the tale have no repercussions during the creation process; for these systems, it is the same to narrate 'The sad moon watched your departure' as it is 'You left.' In contrast, each word selected by a DNN to continue a phrase conditions the future of the narrative.

On the other hand, computational narrative requires the representation of concepts, meanings, and phrases at different levels of abstraction, from the very concrete to the very general. Sentences constitute the concrete part, while structures like those already mentioned form the abstract part. Neural networks operate at the most concise level by linking words and sentences to generate texts, although recent efforts have attempted to build systems that function at more general levels. Thus, important elements that we have studied in previous chapters, such as the representation of complex narrative structures, characters' goals, the protagonist–antagonist relationship, and so on, are not considered by current DNNs but could form part of future combined symbolic/DNN models.

The success and limitations of DNNs have given new life to an old debate about how to integrate the best of symbolic reasoning models with the best of neural models, that is, how to develop neural–symbolic systems. D'Avila Garcez and Lamb (2020) refer to them as the third wave of AI. In the following years, the scientific and technological community interested in the area will focus on studying the fundamental aspects necessary to achieve a greater integration between symbolic reasoning and neural learning. In the words of the authors, 'The key is how to learn representations neurally and make them available for use symbolically (as for example when an AI system is asked to explain itself)' (D'Avila Garcez and Lamb 2020). Marcus (2020) claims that the only way to produce robust artificial intelligence is by developing systems capable of building internal representations of the world that allow these systems to reason about the external world. Marcus underlines the importance of connecting the world of DNN, which focuses on learning, and of classical AI, which focuses on knowledge representation and symbolic reasoning.

These discussions about neural–symbolic systems have led to new considerations, such as what the characteristics of symbolic behaviour in humans are, and how much of them are actually represented in the classical symbolic AI systems. 'Classical perspectives on symbols in AI have mostly overlooked the fact that symbols are fundamentally subjective—they depend on an interpreter (or some interpreters) to create a convention of meaning' (Santoro et al. 2022). Santoro and his colleagues suggest that, rather than developing hybrid models, if DNNs are immersed in human social contexts that demand symbolic behaviour, where people can provide feedback to computer agents, these systems will develop human-like symbolic behaviours. Thus, the confrontation of diverse perspectives on how to improve AI, like those presented here, is essential for the advancement of the field.

The development of neural–symbolic systems will have a great impact on the automatic generation of narratives. In this chapter we describe programs that use basic narrative guidelines such as events (Martin et al 2017), as well as a program inspired by problem-solving techniques (Tambwekary et al. 2019). Although research on how to combine neural and symbolic techniques for automated story generation is promising (see Martin 2021), these systems are still far from achieving the abstract characterizations required for the development of elaborate stories. Going massive does not seem to solve this limitation. As Mark Riedl, a professor from Georgia Tech, put it when talking about GPT-3, 'Making it bigger will just make it model language patterns better. It won't make it a planner or a reasoner or anything else than a language model.'[6] Nevertheless, research on the relationship between the size of a model and its abilities continues (e.g. see Wei et al. 2022). At the same time, other researchers question whether language models are not already too big. These researchers express concerns about environmental costs, financial costs, and problems like the inclusion of under-represented languages in datasets, biases, and so on (Bender et al. 2021).

DNNs are powerful tools that can be used in ingenious ways to generate advanced texts. Novel applications appear every day. For instance, a system trained with a dataset that includes jokes will learn to reproduce them at the right time within a story; if the program is trained with a general indicator like <insert here a joke>, then it will be able to indicate appropriate points in a narrative to insert, for example, computer-generated jokes (Veale 2021). The possibilities are enormous. This chapter only provides a rough idea of what deep neural networks are capable of.

---

[6] Published by @mark_riedl on twitter; accessed 21 July 2020.

11
MEXICA

11.1 The engagement–reflection cognitive account of creative writing

During the second half of the twentieth century, scholars showed an increased interest in the study of cognitive models of writing (for reviews see Becker 2006, Deane et al. 2008, Nystrand 2006). Surprisingly, despite many AI systems based on cognitive architectures that have been developed over the past 40 years (for a review see Kotseruba and Tsotsos 2020), and that few researchers have studied storytelling in humans to inform the development of automatic storytelling systems (e.g. León et al. 2019), we are not aware of any other automatic narrator based primarily on a model of human cognition and writing. In this chapter we introduce the general characteristics of our engagement–reflection cognitive account of creative writing. We explain how this description of the writing process shapes the design of our computer model for narrative generation. Then, we describe MEXICA, a realization of this model, and show an example of how it works.

The engagement–reflection cognitive account of creative writing (Sharples 1999) proposes that creative writing consists of cycles of engagement and reflection guided by constraints. During engagement the writer devotes full attention to creating a chain of associated ideas and turning them into text. During reflection the writer 'sits back' and reviews the material generated so far, invokes memories, transforms ideas, and produces plans and constraints which guide further periods of engagement. Creative writing shows three core features:

1. Writing starts not with a single goal, but with a set of external and internal constraints that include the writer's knowledge and experience, demands of the task, and a primary generator (a key idea that drives creativity).
2. Constraints provide the tacit knowledge to guide the writing process.
3. The constant interchange between engaged writing, guided by tacit constraint, and more deliberate reflection forms the cognitive engine of writing.

Thus, it is the interactions between engagement and reflection that move composition forward. From such interactions emerge different rhythms of work: a writer can switch very quickly between engagement and reflection (e.g. when a writer checks each sentence as it is written) or have long periods of engagement followed by analytical revisions of the entire piece of writing. Those changes in the rhythms of engagement and reflection produce different types of observed writing activity.

An Introduction to Narrative Generators. Rafael Pérez y Pérez and Mike Sharples, Oxford University Press.
© Rafael Pérez y Pérez and Mike Sharples (2023). DOI: 10.1093/oso/9780198876601.003.0011

We employ the engagement–reflection account of creative writing as a theoretical framework for developing our computer model.

11.2 The computer model

Pérez y Pérez (1999) designed the computer model of writing and developed a running prototype called MEXICA. The design of a computer model for narrative generation and the implementation of its prototype is a complex task. It is beyond the scope of this book to present a detailed account of how this process takes place. Rather, we provide descriptions that highlight relevant aspects of it. The chapter offers a general description of the model and its prototype; details about how it works can be found in Pérez y Pérez and Sharples (2001, 2004) and Pérez y Pérez (2007). So knowing that the engagement–reflection cycle is the backbone of the model, it is necessary to find out how to represent its main mechanisms in computer terms. With this purpose, we need to define the following:

- The characteristics of the data structures that represent the writer's knowledge and experience. This is a central component of any computer model. In fact, there is a field in AI known as knowledge representation that focuses on finding the best way to characterize information about the world.
- The details of the processes representing engagement, how they relate to each other, and how they employ the writer's knowledge structures to guide the production of sequences of actions during engagement.
- The details of the processes representing reflection, how they relate to each other, and how they employ the writer's knowledge structures to review and transform the material generated so far, as well as to figure out how to shape the generation of new material.

11.2.1 Engagement

Imagine a story where a person is walking in the forest and suddenly perceives a risky situation, let us say, the presence of a dangerous snake. Probably, this character will react by walking quickly away in order to avoid the predator. To make sense of this description, it is necessary to have some basic knowledge about nature; in this case, one needs to be aware that a poisonous serpent represents a danger for human health and that walking away eliminates that risk. An understanding of the situation makes it possible to determine a logical way to progress the tale. That is, the context in the forest constrains the writer's decision about the behaviour of this hypothetical walker. Thus, our computer model needs a data structure, known as a *story-context*, which represents the writer's knowledge about the current state of affairs of the story-world. You can picture this structure as knowledge in the writer's working memory.

Next, we need to answer an important question. What kind of knowledge should the story-context represent? Herpetology? Animal behaviour? Jogging techniques to run fast? The answer to this question depends on the purpose of the model. In our case, based on a study of the limitations of the knowledge representation employed in previous systems (see the summaries of other programs in this book), inspired by the idea that emotions are the glue of thoughts during the creative process (Gelernter 1994), and by the accepted notion that conflicts are essential elements of any good story (Clayton 1996, 13–15), our computer model represents the story-context as groups of emotional relations and conflicts between characters.

Let us elaborate on this idea. In our computer model, emotional relations represent links between characters expressed in terms of feelings like hate and love. Conflicts represent obstacles that threaten the well-being and intentions of a character. Examples of emotional relations are Character A is attracted to Character B, Character C dislikes Character D, and Character B is enamoured of Character D. Examples of conflictive situations are Character A is chased by a predator, Character A wounds Character B, Character A gets jealous of Character B, and Character A prevents Character B from achieving an important goal. Thus, if José is enamoured of María (this is a story-context that contains one emotional link), he probably will try to perform an action that captures her attention; if Anthony is chased by a snake and therefore his life is at risk (this is a story-context that includes a conflict), he will run. Thus, it is easy to picture an engaged writer, guided by the emotional relations and conflicts that arise as a result of the interaction between characters, producing the following sequence of actions: Juan likes Carmen, Carmen fancies Peter, next Peter also fancies Carmen, so Juan gets jealous of Peter. For a human writer, this sequence of 'what next' actions would be expressed as a flow of text. The following illustrates the content of the story-context for this sequence of events, and how it is updated each time an action is performed. The first action, Juan likes Carmen, establishes an emotional relation from Juan towards Carmen:

Story-context at time = 1
Emotional link 1: Character-Juan is attracted to Character-Carmen.

The second action, Carmen fancies Peter, triggers a new emotional link from Carmen towards Peter; so, the story-context is updated:

Story-context at time = 2
Emotional link 1: Character-Juan is attracted to Character-Carmen.
Emotional link 2: Character-Carmen is attracted to Character-Peter.

In the third action, because Peter responds positively to Carmen's feelings, a new emotional link is triggered:

Story-context at time = 3
Emotional link 1: Character-Juan is attracted to Character-Carmen.

Emotional link 2: Character-Carmen is attracted to Character-Peter.
Emotional link 3: Character-Peter is attracted to Character-Carmen.

As a result, a conflict arises because Juan gets jealous of Peter:

Story-context at time = 4

Emotional link 1: Character-Juan feels attracted to Character-Carmen.
Emotional link 2: Character-Carmen feels attracted to Character-Peter.
Emotional link 3: Character-Peter feels attracted to Character-Carmen.
Conflict 1: Character-Juan gets jealous of Character-Peter

This example illustrates how we picture engagement in the computer model: the constraints provided by the story-context are used by the writer as prompts to select a next associated action to be performed; each time an action is performed the story-context is updated, producing new constraints; these two steps are repeated until an ending condition is met. To be able to represent this process in computer terms, the model requires knowledge about how typical characters react to a given story-context (which we call *contextual-knowledge*) and knowledge about the action each character performs and its consequences (called *story-action*). You can picture these structures as representing general knowledge in the writer's long-term memory.

Contextual-knowledge structures have two parts. The first describes emotional relations and conflicts for a given situation in the story-world. It is similar to the representation of a story-context but rather than employing concrete characters like Carmen, this structure employs variables to represent actors. The second part is a set of possible actions to be performed. For instance, when the life of a character is at risk (e.g. this situation in the story-world might occur when a character finds a dangerous snake), a set of possible next actions to perform are run away, cry for help, or face the threat. Each possible action in a contextual-knowledge structure indexes a story-action.

Contextual-knowledge structure 1

Situation: The life of a character is at risk (a conflict)
Possible actions to perform: run away, cry for help, or face the threat.

We presuppose that this information is part of the writer's knowledge (as we will explain later, this information is obtained from a set of stories written by humans known as *previous stories*).

A story-action represents knowledge about how characters perform. It includes the name of an action, the number of characters participating in the action, and a set of emotional links and conflicts between characters to be triggered when the deed is performed. For example, a consequence of the action 'The snake bites Character A'

Table 11.1 Knowledge employed in MEXICA

Knowledge-structures	Description
Story-context	Represents the writer's current knowledge about the state of affairs of the story-world. You can picture this structure as knowledge in the writer's working memory.
Contextual-knowledge	Represents knowledge about how typical characters react to a given story-context. You can picture this structure as knowledge in the writer's long-term memory.
Story-action	Represents knowledge about the action a typical character performs and its consequences. You can picture this structure as knowledge in the writer's long-term memory.
Dictionary of story-actions	The set of all story-actions.
Previous stories	A set of stories written by humans used to build the contextual-knowledge structures.

is that the health of Character A is at risk (a conflict). The set of all story-actions is known as the *dictionary of story-actions*. Again, we presuppose that this information is part of the writer's knowledge.

Thus, the previous stories and the dictionary of story-actions, along with the contextual-knowledge, form the knowledge-base that represents the writer's prior knowledge and experience. Because the knowledge-base is part of the infrastructure of the system, the model does not represent how this knowledge is created in the writer's mind; it only assumes that it is already there. We explain later how these knowledge structures are built. Table 11.1 shows a summary of all the knowledge the computer model employs.

At this point, we are able to describe the process for unravelling a narrative in terms of computational structures. Given an initial story-context: (1) use this story-context as a cue to probe memory, find a contextual-knowledge structure that is equal or similar to the story-context, and retrieve its set of possible next actions; (2) eliminate from the set all those actions not useful to progressing the text; (3) choose one of the remaining actions in the set as the next action in the narrative, look it up in the dictionary of story-actions, and trigger its consequences in order to update the story-context; (4) return to the first step.

Step (1) describes the process of retrieving a structure that is equal or similar to the story-context. Matching similar structures makes it possible to generate unexpected but still reasonably coherent sequences of actions that do not necessarily correspond to the writer's knowledge. Otherwise, if the system only matches and retrieves contextual-knowledge structures that are identical to story-contexts, the automatic writer would only be able to reproduce familiar situations, that is, it would only be able to 'remember' previous experiences rather that generating novel circumstances not contemplated in its knowledge base. Step (2) removes those possible next actions that are not useful for progressing the narrative. During reflection, the narrative generated so far is evaluated. As a result of this evaluation, a group of guidelines are set. Then, during later engagement, all those actions that do not satisfy the guidelines are classified as not viable and they are eliminated (see later the explanation about guidelines).

11.2.2 Reflection

As we mentioned earlier, during reflection the writer reviews the written material, invokes memories, transforms ideas, and identifies what new material to create. The computer model represents one or more of these activities in the following processes:

- Evaluation of the novelty and interestingness of the material produced during engagement (reviewing material and invoking memories);
- Verification of the coherence of the text produced (reviewing material and transforming ideas);
- Breaking an impasse triggered during engagement (invoking memories and transforming ideas); and
- Generation of guidelines (identifying the features of the new material to create).

The following describes each of these processes:

(a) *Evaluation.* The computer model evaluates the novelty and interestingness of the material produced during engagement. A story is considered novel when it is not similar to any of the previous stories in the knowledge base. A story is considered interesting when the plot includes conflicts and their resolutions.

(b) *Verification of coherence.* During engagement it is possible to generate a sequence of actions that do not flow properly for the tale in progress. In order to solve this problem, *preconditions* are introduced into the model. Preconditions represent the specific common-sense knowledge that needs to be fulfilled before performing an action (this differs from the general situational knowledge contained in the story-context; see Pérez y Pérez (2019) for an analysis of how common-sense knowledge is represented in MEXICA). The function of preconditions is to help the plots flow in a coherent way. For instance, a character can only be healed when she is ill or wounded. So the precondition of the action 'Character A cures Character B' is that Character B must be ill or wounded (a conflict). Story-actions can include a set of preconditions. During reflection, the system verifies that preconditions for all actions in the story in progress are satisfied. When the plot generated during engagement includes unfulfilled preconditions, the system inserts into the tale a story-action whose consequences satisfy the missing requirements. The added action might also have its own conditions that need to be satisfied, which would cause the process to be repeated. Thus, whole episodes can be added to the story to comply with the preconditions of a single action.

The design of computer model also contemplates coherence problems due to spatial location. It is assumed that characters can interact only if they are situated in the same place. Thus, in order to keep the consistency of the story in progress, if the actors participating in an action are situated in different points in the story-world, the system moves one of the characters to the location of the other.

(c) *Breaking an impasse*. Occasionally, human writers get blocked and find it difficult to get ideas to continue a text. Analogously, sometimes during engagement, the story-context cannot match any structure in memory; as a result, the production of material cannot continue and an impasse is declared. To break the impasse, the system analyses the previous stories to identify what events have followed the action that triggered the impasse. For example, if action X triggers an impasse in the tale in progress, and in the previous stories the same action X has been followed by action Y, it is assumed that action Y can also be employed as the next event in the story in progress (of course, preconditions need to be verified). What is expected is that the consequences of performing action Y modify the story-world context in a way that actions can now be retrieved from memory.

(d) *Producing guidelines*. As a result of evaluating the interestingness and novelty of the story in progress, guidelines are set to guide the production of material during engagement. Thus, if the story is too similar to any of the plots in the previous stories, the novelty-guideline becomes active and during engagement the system can only employ actions to progress the tale that have been infrequently used in the previous stories. In this way, using uncommon actions, the model tries to produce original tales. If the story does not include any conflict, the interestingness-guideline is set to boring and during the next engagement the system can only use actions that increase conflicts; on the other hand, if the story includes one or more conflicts but not their resolutions, the interestingness-guideline is set to resolution-required and during engagement the system can only use actions that decrease conflicts (remember that during engagement all possible next actions that do not satisfy the guidelines are eliminated).

Thus, as shown in Figure 11.1, the generation of a story is the result of a constant interplay between engagement and reflection. This process starts with a given initial story-context and then continues until the story is completed or an unbreakable impasse is declared. A story is considered completed when it includes an introduction to a conflict, its development, climax and resolution.

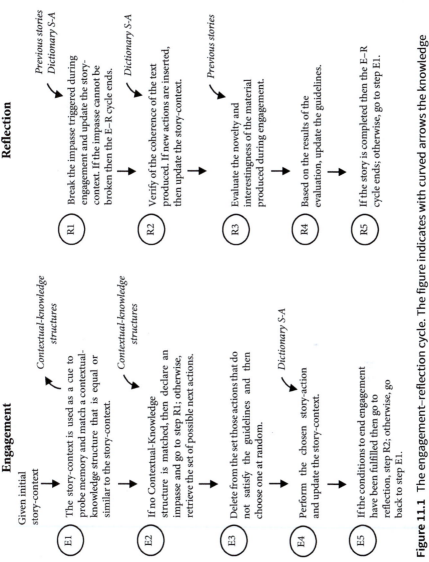

Figure 11.1 The engagement–reflection cycle. The figure indicates with curved arrows the knowledge structures employed at each step

11.3 Instantiating the computer model: the MEXICA program

The computer model is instantiated into a program called MEXICA (it is pronounced 'Meshíca') that produces outline of short stories about the Mexicas, the former inhabitants of what today is México city, also known as Aztecs. The stories created by the program are fictional and they do not have any historical value.

Section 11.2 describes how the theoretical framework has been enriched in order to create the computer model; so the core processes of the system are already defined. Next, we need to establish the details of the infrastructure necessary to run the program. From its conception, MEXICA has been thought of as a research tool. The system offers the user options to test the computer model in different circumstances. The following describes some of these options, which are used to build the infrastructure of the program.

Thus, when MEXICA starts:

1. The program builds its knowledge-structures and keeps them in memory.
2. The user defines all the parameters of the system.
3. The program generates new narratives.

This section describes steps 1 and 2; step 3 is described in the Section 11.4.

11.3.1 The knowledge base: dictionary of story-actions and previous stories

Inspired by the ideas expressed by some writers on how the material written by other authors is a great source of technical knowledge and experiences (e.g. Lodge 1996), a set of tales written by humans—named previous stories, as mentioned earlier—forms the material used to create most of the structures representing the writer's knowledge and experiences. Because in MEXICA stories are characterized as sequences of actions, in order to build the knowledge base the user needs to define a set of valid story-actions and then provide the group of previous stories. Thus, the steps to build the knowledge structures are as follows:

- The user defines the dictionary of story-actions;
- The user defines a set of previous stories; and
- When the program starts, MEXICA builds in memory its knowledge structures through information obtained from the previous stories and the dictionary of story-actions.

The story-actions and previous stories are defined as text files following a syntax designed for that purpose.

(a) *Dictionary of story-actions.* The user must define all the story-actions that MEXICA can use. The definition includes the name of the deed, the number of characters involved, the preconditions and consequences of performing the action—preconditions are optional but all actions must include consequences—and texts in English and Spanish that describe the action; texts are not obligatory and they are used to produce the final version of the story in two languages (although more languages can be included).

The preconditions and consequences included in the dictionary reflect cultural aspects of the type of stories to be generated. Let us elaborate this idea. Preconditions and consequences can be classified as logical and social. The logical ones do not depend on cultural or social contexts. For example, the precondition that a character can only be cured if she is ill or wounded applies to any society. On the other hand, social preconditions and consequences reflect some cultural pattern. For instance, let us imagine an unfriendly society where helping a stranger might not be accepted. So, if we want to reflect this feature in the model, the action 'Character A cures Character B' must include two preconditions: the logical that demands that B must be ill to be cured (a conflict), and the social that requires that A is fond of B to provide some help (an emotional relation).

(b) *Previous stories*. A story is represented as a sequence of actions. The file of previous stories includes several of those sequences. It is assumed that previous stories represent well-constructed stories; that is, they are coherent and include conflicts and their resolutions. Therefore, the structures obtained from them represent correct knowledge. This group of stories is used to build the set of contextual-knowledge structures employed during engagement. As we explained earlier, the previous stories are also employed by the system to break impasses during reflection, so a copy is kept in memory.

The process of building the contextual-knowledge structures works as follows. MEXICA takes the first tale in the file of previous stories and processes each action. Let us imagine that this previous tale includes the following sequence:

1. Eagle knight was impressed with the princess.
2. Eagle knight admitted his passion for her.
3. Eagle knight bought some flowers for the princess.
4. The princess laughed at him.
   . . .

Then the following processes take place:

First, the program performs action 1 and updates the story-context. Let us assume that the consequence of the first action is that Eagle knight fancied the princess. So MEXICA registers in a contextual-knowledge structure that when Character A fancied Character B (this is an emotional link), a logical action to perform is Character A admitted his passion for Character B (action 2 in the tale). Because the contextual-knowledge structure employs variables like Character A and Character B, rather than concrete actors like Eagle knight and the princess, this structure can be employed to develop other narratives.

Next, MEXICA performs action 2 in the tale and updates the story-context. The consequences are that Eagle knight was in love with the princess. So MEXICA registers in a new contextual-knowledge structure that when Character A was in love with Character B (an emotional link), a logical action to perform is Character A bought flowers for Character B (action 3 in the tale).

So, at this point the system has created two knowledge structures:

Contextual-knowledge structure 1
Situation: Character A fancied Character B (emotional relation).
Possible actions to perform: Character A admitted his passion for Character B

Contextual-knowledge structure 2
Situation: Character A was in love with Character B (emotional relation).
Possible actions to perform: Character A bought flowers for Character B

This cycle is repeated until the story ends. All previous stories are processed in the same way.

By the end of this process, MEXICA has produced several contextual-knowledge structures that the program employs to generate new tales. Thus, the features of the knowledge structures in MEXICA depend on the number and content of the previous stories.

The previous stories are written by humans (following a rigid format); therefore, they incorporate social and cultural patterns. Because most of the knowledge in MEXICA comes from the previous stories, the narratives that the program generates might reflect some of those encoded social behaviours.

Recent versions of the system offer mechanisms to outline in a text file known as definitions, the set of available characters and locations in the story-world. By default, the system includes all the actors and locations originally used in MEXICA.

11.3.2 Parameters

A set of parameters that control different aspects of the engagement–reflection cycle can be modified through an interface. For instance, the user:

- Can decide how many actions are generated during engagement before switching to reflection. This parameter makes it possible to reproduce the different rhythms of work described in the theoretical framework, where a writer can switch very quickly between engagement and reflection or have long periods of engagement followed by analytical revisions of the entire piece of writing. By default, this parameter is set to three actions.
- Can specify the percentage of similarity required for the story-context to match a knowledge-contextual structure during engagement. This parameter is closely related to the capacity of originality of the system. If the percentage of similarity is high, the program produces coherent but predictable tales; on the other

hand, if the percentage of similarity is low, the system generates original but chaotic-like outputs that reflection attempts to fix. By default, the percentage of similarity is set to 50%.

- Can deactivate during reflection the generation of guidelines, or even the whole reflective state, to experiment what occurs when the analytical cognitive functions in the model are not working. By default, reflection is set to fully functional.

11.3.3 Establishing the types of emotional relations and conflicts

The program includes a mechanism for defining the emotional relations and some of the conflicts that can be established between characters. By default, the system includes two types of emotional connection: type 1 represents an emotional link between friends that can range from love to hate, and type 2 represents an emotional link between lovers that can range from being in love to hatred (the program represents each type with seven possible values, from −3 to +3, where −3 represents hate and +3 represents love; so, if a character falls in love, the program represents this situation as an emotional link of type 2 and intensity +3).

Similarly, the system includes seven conflicts: when the life of a character is in danger, when the health of a character is in danger, when a character is a prisoner, when a character is murdered, when two characters are in love with the same third character (known as love competition), when a character loves and hates another character (known as clashing emotions), and when a character hates another character and both are in the same location (known as potential danger).

All were selected with MEXICA in mind. However, the user can define up to ten different emotional connections and conflicts that better suit the story theme.

11.3.4 Running the program

When MEXICA starts, the system reads the dictionary of story-actions, the previous stories and the definitions of characters and locations to create its knowledge base of contextual-knowledge structures. Then the user can adjust the parameters that control some relevant features of the engagement–reflection cycle. At this point, the system is ready to generate new stories.

The user must provide to the system an initial story-action (this action can also be chosen at random). MEXICA performs this action, creating an initial story-context, and then the engagement–reflection cycle is triggered. The program offers the possibility of working alone or developing a story in collaboration with the user. MEXICA produces several reports with detailed information about each step completed during story generation (for the sake of clarity, the reports shown in this chapter have been modified).

11.4 How MEXICA works

This section traces the development of the following story generated by MEXICA:

Story 18

Some years ago, the princess was born under the protection of the great god Huitzilopochtli.

The lady was proud to be a member of the Mexica's society.

While she was walking, the princess had a terrible accident and was severely injured.

The lady treated the princess' injuries.

The princess remunerated the lady for all of her help.

Unexpectedly, the lady saw that the princess had the sacred knife that was stolen from the temple. So, there was no doubt: she was the murderer of the old priest!

The princess produced in the lady conflicting feelings.

After consulting a shaman the lady decided to exile the princess.

The lady produced in the princess conflicting feelings. The princess decided to go to the Great Tenochtitlan City. Enraged, the princess provoked and offended the lady.

Quickly, the princess and the lady were immersed in a fight. With all her strength, the lady hurt the princess.

The princess made a potion and drank it quickly. She started to recover.

Endlessly the princess reproached herself for her incongruous behavior.

The princess decided to go to the Quetzalcoatl temple.

The end.

(published in Pérez y Pérez (2017))

The user provides the first action, which is printed in bold:

0 The lady cured the princess

The number on the left side indicates the story time in which the action is generated. In this case, it is assigned a value of 0, since it is the initiating action provided by the user. Then the action is performed, the initial story-context is formed from this first action, and the generation process starts in engagement. As a result of the lady curing the princess, the princess is very grateful towards the lady; so the story-context looks as follows:

Story-context at time = 0

Emotional link 1: The Character-Princess feels very grateful towards the Character-Lady

The program uses the story-context as cue to probe memory and it matches the following two contextual-knowledge structures:

Contextual-knowledge structure 1
Situation: Character-A feels very grateful towards Character-B.
Possible actions to perform: Someone mugged Character-B, Character-A
  rewarded Character-B, Character-A despised Character-B.

Contextual-knowledge structure 2
Situation: Character-A feels very grateful towards Character-B and Character-C
  feels very grateful towards Character-B.
Possible actions to perform: Character-B and Character-A felt attraction for each
  other, Character-B laughed at Character-C.

If in the first structure we substitute Character-A for princess and Character-B for
lady, the contextual-knowledge 1 is the same as the current story-context. That is
why the system matches them. Next, the system retrieves the list of possible actions
to perform and substitutes in the deeds the variables with the appropriate characters:
Character-A for princess and Character-B for lady. The action 'Someone mugged
Character-B' includes an unknown character. MEXICA employs diverse routines to
instantiate unidentified characters (it is beyond the scope of this book to explain how
those routines work; for details, see Pérez y Pérez and Sharples (2001)); in this case,
Someone is substituted with princess because the program attempts to reintroduce
existing characters.

Now, if in the second structure we substitute Character-A with princess and
Character-B with lady, the story-context represents half of the contextual-knowledge
structure 2; the system classifies them as 50% similar and, therefore, they are
matched. Next, the program substitutes Character-A for princes and Character-B
for lady; however, this structure includes a third actor, Character-C. Again, the system employs its routines to instantiate characters and substitutes Character-C for the
actor farmer.

In this way, MEXICA retrieves five possible next actions to progress the tale: the
princess mugged the lady, the princess rewarded the lady, the princess despised the
lady, the lady and the princess felt attraction for each other, and the lady laughed at
the farmer. For MEXICA, all these options are coherent ways of progressing the plot
because all of them have been used as a next action in the same or similar situations
in the previous stories; each of these options would drive the story in different directions. The system selects at random 'the princess rewarded the lady' as the next action
in the tale. Using a random selection allows the system to produce varied and, sometimes, unexpected situations; reflection will try to fix any incoherence that might
arise as a result of this procedure.

After performing the deed, the story-context is updated and the process of retrieving actions during engagement continues. The system generates a sequence of three
actions and then it stops engagement (all actions generated during engagement are
printed in italics):

0 The lady cured the princess
1 The princess rewarded the lady

2 The lady discovered that the princess was a murderer
3 The lady decided to exile the princess

Next, the system switches to reflection and evaluates the tale in progress. After reflection, the story generated so far looks as follows:

4 Some years ago the princess was born
5 The lady was a Mexica
6 The princess had a severe accident
0 The lady cured the princess
1 The princess rewarded the lady
2 The lady discovered that the princess was a murderer
3 The lady decided to exile the princess

Let us explain the processes performed during reflection. First, MEXICA verifies that the lady and the princess are in the same location before performing the action at time 0. So, at times 4 and 5 the system introduces both characters into the story; the consequence of deeds used to introduce actors is to situate characters in the default location, which in this case is Tenochtitlan City (the user can modify the default location).

Next, MEXICA verifies that the preconditions of all the actions are fulfilled and detects that it is necessary to justify why the lady cured the princess. MEXICA looks in the dictionary of story-actions for some deeds whose consequences solve the problem. It finds options like the princess and the lady suffered an accident, the princess had an accident, or the hunter wounded the princess. It selects one at random; so, at time 6, the system inserts the action where the lady suffered an accident and got injured. The system performs the evaluation process and obtains an adequate value for the novelty and interestingness of the tale in progress.

The system switches to engagement but it cannot generate a new sequence of actions; that is, the story-context cannot be matched to any contextual-knowledge structure in memory. An impasse is declared and the program switches back to reflection to try to break the impasse. This is the story generated after reflection:

4 Some years ago the princess was born
5 The lady was a Mexica
6 The princess had a severe accident
0 The lady cured the princess
1 The princess rewarded the lady
2 The lady discovered that the princess was a murderer
3 The lady decided to exile the princess
8 The princess decided to go to Tenochtitlan City
7 The princess insulted the lady
9 The princess fought the lady

MEXICA looks in the previous stories for situations like the one that triggered the deadlock to copy the solution. Thus, at time 7 the program inserts the action where the princess insulted the lady. However, because the princess was exiled, the system detects that she and the lady are not in the same location. Thus, at time 8 MEXICA moves the princess back to Tenochtitlan City. The system performs the evaluation process and obtains an adequate value for the novelty and interestingness of the tale in progress.

The system switches to engagement but, again, a new impasse is declared. The system switches back to reflection and, to try to break the block, at time 9 it inserts a fight between both actors. The system switches back to engagement and this time it is able to generate new actions. This is the story generated so far:

4 Some years ago the princess was born
5 The lady was a Mexica
6 The princess had a severe accident
0 The lady cured the princess
1 The princess rewarded the lady
2 The lady discovered that the princess was a murderer
3 The lady decided to exile the princess
8 The princess decided to go to Tenochtitlan City
7 The princess insulted the lady
9 The princess fought the lady
10 The lady wounded the princess
11 The princess cured herself
12 The princess went to the temple
The end

At time 10 the lady wounded the princess, and at time 11 the princess cured herself. MEXICA can only produce two actions before a new impasse is declared. The system switches back to reflection and at time 12 it inserts the action that moves the princess to the temple.

MEXICA changes to engagement but a new impasse is triggered; it switches to reflection, but this time the system does not find a way to break the impasse and therefore the engagement–reflection cycle ends. At this point, from the eleven actions produced by the system, five were generated during engagement and seven during reflection.

Next, in order to improve the narrative, MEXICA makes explicit in the tale those emotional conflicts that involve clashing emotions:

4 Some years ago the princess was born
5 The lady was a Mexica
6 The princess had a severe accident
0 The lady cured the princess
1 The princess rewarded the lady
2 The lady discovered that the princess was a murderer

13 The princess produced in the lady conflicting feelings.
3 The lady decided to exile the princess
14 The lady produced in the princess conflicting feelings
8 The princess decided to go to Tenochtitlan City
7 The princess insulted the lady
9 The princess fought the lady
10 The lady wounded the princess
11 The princess cured herself
15 The princess reproached herself
12 The princess went to the temple
The end

In the tale, the princess rewarded the lady and as a result the lady was very affectionate towards the princess. However, the lady discovered that the princess was an assassin. This produced in the lady contradictory emotions toward the princess. MEXICA makes explicit this situation at time 13. Similarly, the princess was grateful towards the lady for saving her life, but then she got very angry because the lady exiled her. MEXICA makes explicit this contradictory situation at time 14. Finally, at the end of the tale, the princess felt contradictory feelings towards herself because of her erratic behaviour. MEXICA makes this situation explicit at time 15. The story is finished and the system writes the final version.

The story is realized by employing templates defined by the user in the dictionary of story-actions. Their number and length are not limited, so the same action might be expressed in different ways. The templates include variables that automatically are substituted with characters and pronouns. By default, for each action in the tale, the system chooses at random one of the available descriptions, although it can be forced to only use specific elements. In this way, because these texts are easy to define and modify, different prints of the same story might show different linguistic features.

All the stories produced by MEXICA can be automatically added to the file of previous stories. In this way, the system is able to build new knowledge-structures that come from its own work.

11.5 Self-evaluation and collaborative work in MEXICA

MEXICA performs two different processes of evaluation: one during reflection and one after the final story is generated. During reflection, the evaluation influences development of the narrative to contribute to generating a good story. This process is performed on plots in progress. The goal of the evaluation of the final product is to assess whether the generated story as a whole is coherent, novel, and interesting (Pérez y Pérez 2014). This process is performed on concluded works. As far as we know, MEXICA is the only automatic storyteller that assesses its own works. As part of the evaluation of the final story the system verifies the following:

- The plot introduces a conflict at the beginning of the tale, that the conflict is developed until it reaches a climax, and that at the end the conflict is resolved.

- The new story is novel; that is, it produces story-contexts not present in its knowledge base. In this way, the system can learn from its own stories.
- The story does not include repeated sequences of actions.
- All actions serve a purpose in the plot, such as locating characters in the same place. MEXICA eliminates all those useless actions to produce a better story.

MEXICA analyses the final story and calculates a score. Each time one of these characteristics is not fulfilled the score is decremented. The following shows the report of the evaluation of Story 18:

Evaluation of Story 18

This is a good story. Here are some comments about this work.

- Coherence

The story is very logical; all actions are nicely integrated and form a coherent unit.
The opening is good.
It has a clear closure where all conflicts are worked out. Nice!

- Interestingness

The text has a good introduction.
The story reaches a nice climax with a good amount of tension.
The enjoyable closure of the story makes the reader to connect nicely with the plot.
There are surprising events that make the story appealing. I enjoyed that!

- Novelty

I find this story pretty original! I love it!

- Score

The evaluation of your story is 100/100

11.5.1 MEXICA-impro

MEXICA is able to produce collaborative narratives with other MEXICAS; we refer to this system as MEXICA-impro (Pérez y Pérez 2015b, 2015c). The goal of this project is to study how our cognitive architecture performs in a social environment. It works as follows:

- There are two MEXICA programs working independently that communicate with each other employing a blackboard architecture (the blackboard is a text file that both programs can read and modify at any time).
- Each MEXICA must have its own knowledge base; for example, they might have a different set of previous stories.

- MEXICA-1 starts the process, runs one engagement–reflection cycle, and writes the resulting sequence of actions as text on the blackboard.
- MEXICA-2 reads the blackboard, interprets the story in progress (i.e. it updates its own story-context), runs one engagement–reflection cycle to progress the plot, and writes the resulting tale as text on the blackboard.
- MEXICA-1 reads the blackboard, interprets the story in progress (i.e. it updates its own story-context), runs one engagement–reflection cycle to progress the plot, and writes the resulting tale on the blackboard.

This process continues until the story is finished or an unbreakable impasse is declared. None of the systems can modify what the other program wrote. In this way, because they have a different knowledge base, MEXICA-impro makes it possible to explore how automatic storytellers need to reach agreements in order to collaborate.

11.6 Final remarks about MEXICA

MEXICA is a computer model based on the engagement–reflection cognitive account of creative writing. During the engagement, the system produces material driven by content and rhetorical constraints avoiding the use of explicit goal-states or story-structure information. During the reflection, the system breaks impasses generated during engagement, satisfies coherence requirements, and evaluates the novelty and interestingness of the story in progress. If the results of the evaluation are not satisfactory, MEXICA can modify the constraints that drive the production of material during engagement. In this way, the stories produced by the program are the result of the interaction between engagement and reflection. Once the story is finished, MEXICA assesses the final product in order to rank its own work. The stories produced by the system can be automatically incorporated into its knowledge base. Some relevant features of MEXICA are the following:

- This project demonstrates the computational plausibility of the engagement–reflection cognitive account of creative writing. This approach offers an alternative to the predominant views of writing as a problem-solving activity (see Chapters 3 to 7) or, more recently, as a deep neural network architecture (see Chapter 10).
- The design of the computer model and the development of the prototype results in the enrichment of the theoretical framework. In this way, this project results in detailed explanations of the operation of engagement–reflection cycle and the types of knowledge structures required for the development of the prototype.
- The use of emotional relations and conflicts between characters to represent knowledge structures provides flexibility during the generation of a tale. In this way, predefined story-structures and goal-states are avoided.

- Because stories are the result of the interaction between engagement and reflection, events are not produced linearly. The story emerges as a result of the interaction between the multiple components of the system at different times.
- The use of the dictionary of story-actions, previous stories, and definitions provides the user with a relatively simple way of controlling the system's knowledge-structures.
- The capacity to evaluate its own stories and incorporate them automatically as part of the previous stories gives MEXICA a means to monitor and improve its performance.
- MEXICA-impro generates collaborative narratives, providing a space for the study of some social aspects of automatic narrative generation.

The program exhibits limitations. In MEXICA, some story conflicts result from an inference process, rather than being triggered by a story-action's consequences. For example, to sense a tension due to love competition between actors, the program must constantly check whether any two characters in the tale are both in love with the same person. This mechanism is part of the code of the program. Therefore, the user cannot add new inferred conflicts. Something similar happens with the guidelines; although they are flexible, their number and type are fixed in the code. Because guidelines are an essential part of engagement, this situation might prevent the system from generating stories in a different context.

The computer model of engagement and reflection only attempts to represent a small fraction of the multiple processes that humans perform during writing. It is necessary to increment its cognitive capacities. There are language models that describe the structure or meaning of language a writer uses; world models, of a writer's knowledge of settings and characters; process models of how ideas are formed and how the writer plans, drafts, and revises. This chapter has focused on a cognitive process model, but a full narrative generator also needs to have (at least) good language and world models. Similarly, it would be useful to incorporate the possibility of generating texts rather than employing templates. Thus, this is only the first step in a drawn-out path. MEXICA is a long-term project that is in constant development. Hopefully, it will provide us with interesting surprises.

12

Endless ways to use computers to study narrative generation (part 1)

12.1 Studying computer-based narrative generation from different angles

The purpose of this and Chapter 13 is to illustrate how computers offer endless ways of approaching the study of narrative generation. Thus, rather than providing an exhaustive review of story-generator systems, these two chapters describes seven computer programs that exemplify the diversity of research interests and technical possibilities, in order to represent in computer terms some core aspects of narrative generation not covered earlier in this book.

12.2 MESSY

Sheldon Klein and his students developed the Meta-symbolic simulation system (MESSY) (Appelbaum 1976), which includes two main components: (1) a behavioural simulation programming language that makes it possible to model, generate, and manipulate sequences of events that change through time in a story-world; and (2) a simulation system that executes the instructions specified through the language. You can picture MESSY as a programming language designed to produce stories. These are its main components:

- *Representing the story-world*. MESSY represents the story-world in terms of relations between entities of the type <Subject—Relation—Object>; for example, the relation <man—has—bicycle> represents the assertion that a man has a bicycle, and <Elton—loves—Laura> represents that Elton is in love with Laura. Entities and relations might be connected to other entities and relations; in this way, it is possible to represent elaborate situations as networks of interlinked subjects and objects. The programming language employs instructions, known as actions, to create, delete, or modify the story-world.
- *Simulated time*. MESSY includes an internal clock that controls the story-time inside the simulator. Time units can be specified as weeks, days, hours, and minutes. When required, the system increments the story-time. When the

An Introduction to Narrative Generators. Rafael Pérez y Pérez and Mike Sharples, Oxford University Press.
© Rafael Pérez y Pérez and Mike Sharples (2023). DOI: 10.1093/oso/9780198876601.003.0012

story-time reaches a stipulated finishing time the simulation ends (the finishing time is defined by the programmer). In this way, it is possible to design a narrative that lasts minutes, days, or weeks.

- *Actions.* Actions are instructions defined as part of the language. They are used to modify the story-world, instantiate variables, generate random numbers, enable or disable groups (see later explanation), end the simulation, and so on.
- *Rules.* Rules are structures that comprise a list of actions to be performed and a set of preconditions that need to be fulfilled to trigger the rule. Because actions can modify the story-world, rules are useful for controlling the behaviour of characters and for producing ad hoc situations for progressing a tale. Preconditions usually represent facts that must be true in the story-world, but they might also include probabilistic elements. For instance, a rule might include the following precondition: if character W is reasonably attractive, there is a 45% chance that this rule will be executed; otherwise, there is only 5% chance.
- *Groups.* Groups are structures that comprise any number of rules; the rules in each group are performed sequentially. Each group has an associated story-time that indicates when it is next scheduled for execution. MESSY has a table that puts together all those groups scheduled to be performed at the same story-time; when of them have been executed, the system increments the story-time by one unit and repeats the process. Thus, if a group is programmed to be effected every twenty-four hours, and the simulation lasts more than a day, it will be performed once every story-day. If a group is designed to represent events that happened during breakfast, the simulator ignores this group when the story is taking place in the evening and considers it again the following morning. Groups might also be intentionally disabled or enabled by the programmer. In this way, it is possible to prevent a group designed to set the right context at the beginning of a tale from being executed again later.

In MESSY, a program is divided into sections:

- $LIMITS. In this section, the initial values for the internal clock are provided.
- $NODES. In this section, all the entities (subjects and objects) that will be used by the simulator are defined. In MESSY, entities are referred to as nodes. Some examples of nodes are characters like Dr. Hume, John, Lord Edward, Lady Buxley, Lady Jane, and Marion; locations like park, hotel, tennis court; objects like telephone.
- $RELATIONS. In this section, the programmer defines all possible relations. Some examples are affection, blackmail, caress, gossip, attractive, wealth, jealous, and married.
- $CLASSES. The programming language in MESSY makes it possible to define groups of entities (or nodes), known as classes. Some examples of classes are (Female = Lady Buxley, Lady Jane, Marion), (Male = Dr. Hume, John, Lord Edward), (Locations = park, hotel, tennis courts).

- $NETWORK. In this section, the programmer indicates the initial state of the story-world. Some examples are Lady Jane is wealthy, Lady Buxley is very attractive, and Lady Jane feels affection for Lord Edward.
- $Group. In this section, all groups and their associated rules and actions are defined. The following description illustrates the definition of a group. It is adapted from an example provided by the authors. This group is designed to represent a situation where a character discovers that two other persons are having an affair; depending on who discovers the affair, the plot might give rise to blackmail, gossip, or a jealousy situation. It employs the variables X, Y, and Z which are instantiated with characters.

> Group: Discovering the affair. This group is scheduled to be executed when the story-time is equal to 10 minutes.
> Rule 1: The system assigns to the variables X and Y a male and a female character that will have an affair.
> Rule 2: The system chooses at random one location, places X and Y close to each other in that location; they caress and they are lovers.
> Rule 3: The system assigns to the variable Z a random female character that is different from the one having the affair and who discovers the situation between X and Y.
> Rule 4: Character Z will blackmail Y as long as the following conditions are satisfied:
> - Z is not the detective or the spouse of any of the characters having the affair.
> - Y is married.
> - Z is not already blackmailing Y.
> - Z needs money.
> Rule 5: If Z is the spouse of one of the people having the affair, then Z will get very jealous.
> Rule 6: If Z does not blackmail anyone or get jealous, and if X is married, then Z will gossip with the spouse of X about the affair.

- $END. This command indicates the end of the program.

When MESSY starts, the following steps are performed:

1. The internal clock is initialised
2. MESSY executes the rules in all the groups scheduled to the current story-time.
3. If the story-time is equal to the finishing time, or the action to end the simulation is performed, or no group can be executed, then MESSY stops.
4. Otherwise, the story-time is incremented by one unit and then the system goes back to step 2.

Employing MESSY, a murder story that took place during a house party was designed (Klein et al. 1973). The authors employed a grammar that translates the story-world representations into English. Thus, the narrative comes from the successive reports about how the story-world changes each time the story-time is updated. The following is an extract from this story (at this moment in the plot, the characters James and Marion are married):

> The day was Monday.
> The pleasant weather was sunny.
> Lady Buxley was in a park.
> James ran into Lady Buxley.
> James talked with Lady Buxley.
> Lady Buxley flirted with James.
> James invited Lady Buxley.
> James liked Lady Buxley.
> Lady Buxley liked James.
> Lady Buxley was with James in a hotel.
> Lady Buxley was near James.
> James caressed Lady Buxley with passion.
> James was Lady Buxley lover.
> Marion following them saw the affair.
> Marion saw the affair.
> Marion was jealous.
>
> (Published in Klein et al. (1973))

As you probably have noticed, this scene was created using the group we described earlier. The system employs the same group several times during the story: on Tuesday, Marion and Dr Hume become lovers and Lady Jane blackmails Marion; on Wednesday, Lady Jane and John Buxley become lovers and Cathy blackmails Jane; and so on. The whole story is 2100 words.

We would like to stress the fact that this work was undertaken in the 1970s; it represents one of the first attempts to build an automatic narrative generator.

12.3 SCÉALEXTRIC

SCÉALEXTRIC (Veale 2017) is a system for narrative generation that employs repositories containing diverse data, carefully crafted by the author, which is combined to develop stories. The system performs three main processes: development of the plot, selection of the main characters and their ancillary figures, and realization of the plot.

12.3.1 Development of the plot

SCÉALEXTRIC employs a repository known as 'script midpoints' that includes more than 2,700 structures, each one made of sequences of three actions, which work as building blocks during plot generation. The succession 'A are-flattered-by B, A trust B,

A are-manipulated-by B' illustrates one of those structures, where A and B represent characters that are instantiated in a posterior process.

The system employs two different ways to progress the plot; we have named them the *left-to-right* and *top-down* techniques. The system progresses a plot in a left-to-right fashion by attaching new triplets to a given sequence of actions. The addition must fulfil the following condition: the last action of the current sequence must be equal to the first action of the triplet to be attached. For instance, given 'A is-flattered-by B, A trusts B, A is-manipulated-by B' the system looks in the repository script midpoints for a three-action sequence that starts with 'is-manipulated-by'; the following structures are the result of this search:

- 'A is-manipulated-by B, A stands-up-to B, A is-promoted-by B'
- 'A is-manipulated-by B, A grows-suspicious-of B, A stands-up-to B'
- 'A is-manipulated-by B, A criticizes B, A is-denounced-by B'

The program selects one of them at random. If the last option is chosen, the resulting plot looks as follows: 'A is-flattered-by B, A trusts B, A is-manipulated-by B, A criticizes B, A is-denounced-by B'. The event connecting the two sequences is only represented once. Actions cannot be repeated along the narrative to avoid getting trapped in a loop. SCÉALEXTRIC finishes the plot when a predefined number of actions has been reached.

For the second approach, rather than appending new events, the program *expands* a triplet, inserting actions between the verbs that compose the sequence. A set of rules specifies how to perform this process. For instance, 'A reports-to B, A spies-on B, A is-reported-by B' has associated with it the following rules:

Rule 1: The pair of actions 'A reports-to B' and 'A spies-on B' can be expanded by inserting in the middle any of the following events: 'A is-disgusted-by B', 'A is-shocked-by B', 'A is-sickened-by B', 'A turns-against B'.

Rule 2: The pair of actions 'A spies-on B' and 'A is-reported-by B' can be expanded by inserting in the middle any of the following events: 'A is-refused-payment-by B', 'B withholds-payment-from A'.

Rule 3: The pair of actions 'A spies-on B' and 'A is-refused-payment-by B' can be expanded by inserting in the middle the following event: 'A blackmails B'.

A rule might provide several options to expand the triplet. Thus, an enormous number of different scenarios might be generated. Here is an example. If given the series 'A reports-to B, A spies-on B, A is-reported-by B', the system triggers Rule 1; this sequence might be expanded as 'A reports-to B, **A is-shocked-by B**, A spies-on B, A is-reported-by B' (the inserted action is marked in bold). If the system then triggers Rule 2, the sequence might be expanded as 'A reports-to B, A is-shocked-by B, A spies-on B, **A is-refused-payment-by B**, A is-reported-by B'. If the system then triggers Rule 3, the sequence is expanded as 'A reports-to B, A is-shocked-by B, A spies-on B, **A blackmails B**, A is-refused-payment-by B, A is-reported-by B'. Because no more rules can be applied the story is finished.

12.3.2 Choosing the core characters

SCÉALEXTRIC employs a repository known as the non-official characterization list (NOC) (Veale 2016), which includes as many as twenty-six fields for more than 1,000 real and fictional characters. This is a partial example of the NOC's data-structure:

> Description-of-Character:
> Name:
> Portrayed-by:
> Category:
> Fictive status:
> Negative talking points:
> Positive talking points:
>
> …

The system employs the NOC list to find the best two characters to instantiate the plot generated in step (1). The first character must suit the initial incidents in the plot. To achieve this goal, the system includes a set of directions that links initial actions with characters' categories included in the NOC list. Here are some examples. Let us imagine a plot that starts with the following sequence 'A is-elected-by B, A disappoints B, A is-despised-by B'. The system includes the following order:

> When the plot starts with 'A is-elected-by B' or 'A appoints B' or 'A is-targeted-by B' or 'A campaigns-against B' or 'A runs-against B' or 'A votes-against B', then look in the NOC list for a character that is categorized as 'politician'.

The NOC list comprises information about real and fictional politicians like Nelson Mandela, Neville Chamberlain, Polonius, Condoleezza Rice, and Margaret Thatcher. The system chooses one of them at random. For this example, let us instantiate character A with Margaret Thatcher. Next, the system looks for a good match for her. There are multiple ways in which the attributes in the NOC list can be used for this purpose. For instance, because Thatcher was a real prime minister the system might look for a fictional politician to instantiate character B. You can imagine a story where Thatcher and Polonius act together. One of the most powerful features of fiction is the possibility of creating a setting where a historical and a nonexistent character interact; the information encoded in the SCÉALEXTRIC's knowledge-base allows for this possibility.

The NOC list can be exploited in more ingenious ways. One of its attributes registers the name of the actor that portrayed a real or fictional character in a movie, on TV or stage. For instance, Meryl Streep portrayed Margaret Thatcher in the film *The Iron Lady* and the chef Julia Child in the film *Julie & Julia*. Thus, Streep can be used as an indirect link between a primer minister and a chef that worked as a TV host. Through this association the system can produce a singular narrative where Thatcher and Child are the core protagonists.

The current version of SCÉALEXTRIC includes about ten or so explicit instructions about how to combine individuals, known as schemas. However, the NOC list offers an enormous number of possibilities for pairing characters in surprising ways.

12.3.3 Realization

SCÉALEXTRIC realizes a story using templates. The system employs three main types of predefined texts: those that express how an action is performed, those that provide an opening for a story, and those that properly close a narrative.

Each action in the system is associated with two or more texts that express in English the way an event is performed by characters. For instance, the deed 'A is-manipulated-by B' can be rendered as 'B pulled A's strings', 'B knew how to push A's buttons', or 'B knew just how to manipulate A'. The deed 'A disappoints B' can be rendered as 'A thoroughly disappointed B', 'B thought "What a loser" when looking at A', or '"Could you be a bigger disappointment?" asked B sarcastically'. Thus, for each action in the plot the system selects at random one of the associated texts to realize the plot.

Next, the program chooses an opening. Actions are linked to texts designed to properly start a narrative. For instance, if the plot begins with the action 'A disappoints B' any of the following texts can open the tale: 'A was one of life's losers', 'Disappointment was a gift that A gave to others', or 'Failure followed A the way flatulence follows a curry'. Thus, if a plot starts with the sequence 'A disappoints B, A is-resented-by B, A is-stalked-by B', the opening and the first action of this story might be rendered in English as 'Failure followed A the way flatulence follows a curry. "Could you be a bigger disappointment?" asked B sarcastically . . .'.

Similarly, actions are linked to texts designed to end a tale. If a story ends with the action 'A is-denounced-by B', the system can include as an epilogue any of the following texts: 'Thereafter A crawled off to lick its wounds in private', 'Thereafter A crawled off to plot its revenge in the shadows', 'A was down for the count but would soon rise again'.

The system also includes explicit information about how to link two events in the plot using connectors like 'so', 'then', 'and', or 'but'.

In summary, SCÉALEXTRIC works as follows:

1. The system chooses at random an initial three-action sequence from the repository 'script midpoints'.
2. The system develops the plot using either the left-to-right or top-down technique.
3. The system looks in the NOC list for those characters that best suit the plot.
4. The system instantiates characters and then renders the sequence of actions that compose the plot; finally, it finds an appropriate introduction and epilogue.

The following is a narrative generated by SCÉALEXTRIC that Veale provided to us for this book.

What if Malcolm X hired Robin Hood?

Malcolm X saw critics as a necessary evil that one must tolerate. So at first, Robin reviewed Malcolm's work to form a critique.

'DISAPPOINTED!' shouted socially conscious Robin when Malcolm set about giving to the poor.

So Robin fired Malcolm for being lousy at giving to the poor.

Because earlier Malcolm leaked valuable information to Fletcher Christian.

But because of this Fletcher gave Malcolm a job.

Whereupon Malcolm designed beautiful things for Fletcher.

So in due course Fletcher made a lot of money from Malcolm.

After which Fletcher funded Malcolm's business ventures.

Thus funded, Malcolm formed a rivalry with Robin as to who was best at giving to the poor.

Up until this point King Henry VIII funded Robin's business ventures.

But now Fletcher spread slander about Robin.

At this low point Henry lost all faith in Robin.

So, to no one's surprise, Henry and Robin went different ways.

Moved by this turn, Robin reached out to make a connection to Malcolm.

For many reasons Malcolm owed Robin a great deal.

So, it came to pass, Robin made a heartful appeal to Malcolm.

Following this appeal Malcolm hired Robin to help him with fighting for civil rights.

So Robin erected buildings for Malcolm.

Then Malcolm offered an honest review of Robin.

In the end, Robin's plight really moved Malcolm.

Thereafter Malcolm could do no harm to Robin but felt a common cause with Robin.

SCÉALEXTRIC employs twelve carefully designed repositories that connect their elements. These relations represent knowledge provided by the designer of the system in a painstaking fashion that makes it possible to combine in unforeseen ways the different components that compose the knowledge-base. Because at each point of decision the program provides multiple options, SCÉALEXTRIC is able to generate an enormous number of different plots. All repositories can be downloaded as spreadsheets and adapted for use by any user.[1] If the data in the knowledge-base are modified, then the system adapts to the new content. This is known as soft-coded rules. In other words, soft-coded rules are built into the knowledge structures that the program uses; you can change these without changing the program.

---

[1] https://github.com/prosecconetwork/Scealextric

12.4 GRIOT

GRIOT (Harrell 2005) is a platform for generating interactive narratives. These interactive narratives can be multimedia and/or text-based. Here we focus only on production of text. The author of a work implemented using GRIOT typically carefully plans the piece in advance (although it is also possible to approach more improvisationally, especially if building on a prior example); she must also either define the content of the data structures that the program requires or use content from pre-existing data structures. Works in GRIOT can alter with each execution and retain coherent meaning and style despite those changes. For instance, the plot of a narrative work may be fixed, but some elements of the text may be different on each execution. These differences can be driven interactively by user input.

To generate figurative language, the system employs the Alloy algorithm (for details see Goguen and Harrell 2009; Harrell 2007), which creates blends of mathematical structures called semiotic spaces that represent concepts. Because this algorithm produces output in a logical form, a separate function produces natural language output from this. Alloy is based on the principles of conceptual blending theory (Fauconnier and Turner 2002; Turner 2015). In this way, GRIOT employs Alloy to produce expressions which enrich the original text.

In the following, we provide a general description about how GRIOT works. Before starting, the designer of a piece must perform the following activities:

- Write a text (e.g. story or poem) about a relevant topic. This story is represented as a sequence of texts that will be employed to create templates; templates could be considered the main building blocks of the system.
- Define the themes. The designer must specify one or more sets of concepts—implemented as data structures—typically these represent themes related to the topic of the story. Their purpose is to provide the information that Alloy requires to work. Each theme has a name and consists of several data, such as a set of data types, a set of constants of those data types, or a set of binary relations between constants. Binary relations represent relevant concepts for the piece, such as the concept that youth causes envy is represented as causes (youth, envy), or the idea that a girl uses balm is represented as uses (girl, balm). Thus, when GRIOT is running and two themes are chosen (we will explain later how they are chosen), the system selects at random one concept from each theme, and uses them as input to Alloy.
- Define the templates. Templates have a fixed text and they might also include a variable known as 'wildcard' that will be replaced by content generated by the Alloy algorithm; this replacement process is known as instantiation. These wildcards specify the grammatical form that the output of Alloy must be mapped to by the natural language function. Optionally, a template might also indicate a specific theme that should be used for the instantiation process. An example is 'This is the fixed text (* g-singular-noun d-nameTheme)'. The

parentheses and the star indicate that this is a wildcard. The expression g-singular-noun ('g' stands for generated content) specifies the grammatical form in which the output of Alloy must be expressed. The expression d-nameTheme ('d' stands for domain, referring to the set of concepts that can be used to represent a theme) denotes that a theme named nameTheme must be used as one of the inputs to Alloy. When the second parameter is missing the system chooses one theme at random.

- Define the sections of the piece. The narrative to be produced must be divided in what in this book we refer to as sections, each with a name. Sections can also be divided into subsections. Then the user must specify which templates belong to each section.
- Establish the narrative structure of the piece. GRIOT has mechanisms for implementing complicated narrative structures. Let us elaborate this idea. Imagine a piece that includes three sections, Section 1, Section 2, and Section 3. A possible structure is

Section 1 → Section 2 → Section 3

The system takes one template from Section 1, one from Section 2, one from Section 3, and then the process ends. That is, the output is the sequence of these three templates. Because each section typically has many templates to choose from, GRIOT can be set to avoid repeating templates, and the content of templates can change via generated wildcard replacement; there is a high probability that each time the system runs, the output changes.

Sections can be split into two or more subsections. For instance, Section 2 can be divided into Subsections 2a and 2b. Each subsection is a nested element of the main section. Thus, it is possible to combine sections and their nested elements:

Section 1 → Subsection 2 → Subsection 2a → Subsection 2b → Section 3

The program provides ways that make it possible to repeat a section, a sequence of sections, or a section and their nested subsections several times, such as

Section 1 → Section 2 → Section 2 → Section 3

Section 1 → Section 2 → Section 3 → Section 2 → Section 3

Section 1 → Subsection 2 → Subsection 2a → Subsection 2b → Section 3 → Subsection 2 → Subsection 2a → Subsection 2b → Section 3

Also interesting, GRIOT allows for a section or a sequence of sections to be repeated a bounded random number of times. The designer must specify two numbers, and then the system calculates a random value between those numbers. In the following example, the line over Sections 2 and 3, and the numbers 1 and 4 over the line, represent a loop where such a sequence will be repeated between one and four times during the generation of the piece:

1_____4
Section 1 → Section 2 → Section 3 → Section 1

The previous example might be expanded as follows (with Section 2 → Section 3 repeated twice):

Section 1 → Section 2 → Section 3 → Section 2 → Section 3 → Section 1

For all of these reasons, the number of combinations GRIOT can generate is enormous.

As part of the narrative structure, the designer must also specify when the system will interact with the user. In the following example, we employ a letter R (that means 'read a theme word from the user') to indicate those sections where the interaction occurs:

1_____4
Section 1 → Section 2^R → Section 3^R → Section 1

Thus, each time that Section 2 or 3 starts, the user must introduce the name of one of the themes that the designer defined for the piece. This theme is used during the process of instantiating a template.

Once these elements are in place, the designer can make use of GRIOT. The system is programmed in Lisp. The user is required to type into the original code of the system all the information described earlier employing the Lisp notation.

The following shows an example of a piece from the project *Living Liberia Fabric*. This work is a multimedia memorial developed to support peace after years of civil war in Liberia (for details see Harrell (2013), 99–110). Here we only focus on the textual part.

Woman-focused poetic narrative
our lost mother reflecting
not a story, a human
the first Americo-Liberians acquired land with arms
there were positive stories of survival
Margibi 3,394 victims
that woman forgetting there
Rivercess 2,315 victims
what can we do in the future?
regal, everyday women lead the way
it is a process driven by the people
the post-colonial often recaps the venom of the colony
woman forgetting, we can be one
we can love again

look toward ourselves
possibility

(Published in Harrell (2013), 105)

To illustrate how GRIOT works, we describe how the first five lines of *Living Liberia Fabric* are generated. This piece includes thirteen sections. In this example we only employ five: level-0-orientation, level-0-to-1-transition, level-1-orientation, level-1-primary-narrative, level-1-background-material. Table 12.1 shows examples of templates associated to each section.

Although this piece includes twelve themes, for this example we only mention two of them, the theme Woman and the theme Memory. Woman includes relations like is (woman, surviving), is (lady, mother). Memory includes relations like is (dreaming, reflecting), is (forgetting, losing).

The expanded narrative structure necessary to produce the first five lines of the piece is

level-0-orientation^R → level-0-to-1-transition^R → level-1-orientation^R →
    level-1-primary-narrative^R → level-1-background-material → . . .

Now, the GRIOT-based Living Liberia Fabric starts. It works as follows.

(1) Based on the narrative structure of the piece, the system first processes the section level-0-orientation^R. The superscript ^{R} indicates that the user should type into the system one of the thematic words of the piece. In this case, the user chooses Woman. Next, the system selects at random a template from those associated with the section level-0-orientation (see Table 12.1). In this case, the system chooses the template 'our lost (∗ g-singular-noun d-memory)'. This template needs to be instantiated by the Alloy algorithm by generating a singular noun. Alloy requires two themes as input; the first is Woman, provided by the user to replace g-singular-noun, and the second is Memory, as it is indicated by the parameter d-memory in the template. Then, the system selects at random two concepts, one from each of these themes: is (lady, mother) from the Woman theme and is (dreaming, reflecting) from the Memory theme. These are the inputs to Alloy; as output it delivers the text 'mother reflecting', which represents the blending of the input concepts. In this way, the instantiated template 'our lost mother reflecting' becomes the first sentence in the piece.

(2) Next, the narrative structure requires processing the section level-0-to-1-transition^R. The same process is repeated. The system selects the template 'not a story, a human' from those associated with the section level-0-to-1-transition. Because it does not include a wildcard, no more processes are necessary. In this way, the second line of the piece has been generated.

(3) The system repeats the same process for lines three to five. In each case, a template without wildcard is selected from the available options, and in this way the sentences 'the first Americo-Liberians acquired land with arms', 'there were positive stories of survival', and 'Margibi 3,394 victims' are generated.

Table 12.1 Some sections of the piece Living Liberia Fabric and their associated templates

Section	Templates
level-0-orientation	• a silent moment for the lost • one of 1.5 million displaced • displaced (∗ g-singular-noun d-memory) • our lost (∗ g-singular-noun d-memory)
level-0-to-1-transition	• tracing the tangled roots of the loss • everyday survival, (∗ g-singular-noun d-memory), hero • not a victim, a brother • not a story, a human
level-1-orientation	• the modern nation of Liberia, as it exists today, was partly shaped by the transatlantic slave trade • the early Americo-Liberians dominated the security sector • commerce in the early days was dominated by the Americo-Liberians • the first Americo-Liberians acquired land with arms
level-1-primary-narrative	• we cannot map out the guilt in the conflict • there were positive stories of survival • women provided everyday heroics in the face of atrocity • there were everyday heroes in the face of tragedy
level-1-background-material	• Margibi 3,394 victims • Sinoe 5,706 victims • Rivercess 2,315 victims • Maryland 3,934 victims

GRIOT is not just a platform for the interactive generation of narrative, but a program for exploring the expressivity, subjectivity, and cultures of poetic language and meaning.

12.5 A chess-game narrator

Gervás (2012, 2014) developed a system that tells the actions in a chess game. The program reads a description, written in algebraic notation, of all movements in a given game, for example, 'move 1: e4 c5, move 2: Nf3 d6, move 3: d4 cxd4 . . ', and uses this information to output a narrative. The board is seen as a model of the story-world in which events unfold. Chess pieces represent the characters that take part in the story. Each character only 'perceives' what happens in the squares surrounding it; the number of nearby squares that an actor can sense is a modifiable parameter. As a result, the system is able to relate the happenings on the board from different angles. If a white pawn captures a black pawn, the scene can be recounted from the point of view of the white piece, the black piece, or some other piece close to the action. The challenge is to select a coherent telling from all the possible events and perspectives generated during a game. To achieve this goal the system must decide which events to narrate and which to omit, their order of presentation, and the character that the narrative follows (called the focalizer).

The program employs templates for producing the final output. Each chess movement corresponds to one story-day, so the program is able to employ temporal expression like 'one week ago' or 'one month earlier' in its output. The position of all pieces is registered at all times, allowing the system to include spatial information in the output.

The future of all participants in the game are known in advance; the system exploits this information to narrate fatal confrontations between two characters, interleave the 'life' of different characters into a single narrative or search for all possible stories that can emerge from a given game. The following example shows a confrontation between the white and black queens. To produce this narrative the system gathers together the 'life' of the black queen from the beginning of the game until the confrontation starts. The system then includes the life of the white queen from the beginning to the same point. Finally, the system adds the result of the attack and the fate of the black queen. This is an excerpt from the narrative:

> The black queen was four squares north of the centre of the board. The third black pawn was to the right.
>
> (...) The black queen saw the third black pawn leaving to the right.
> (...) Three days later, the black queen moved southeast. The third white pawn remained behind.
>
> (...) The black queen saw the white queen appearing ahead. The black queen attacked the white queen. A month earlier three squares northwest, the white queen was three squares south of the centre of the board.
>
> (...) The white queen saw the black queen arriving. The black queen attacked the white queen.
>
> The white queen died. The black queen saw the white right bishop arriving. The white right bishop attacked the black queen. The black queen died.
>
> (published in Gervás (2012))

The first line introduces the black queen and her location. The sentence 'The black queen saw the third black pawn leaving to the right' indicates that the black pawn has moved away from the black queen's perception. The sentences 'The black queen saw the white queen appearing ahead' and 'The white queen saw the black queen arriving' narrate two different perspectives of the same action.

12.6 Final remarks about the four narrative generators

In this chapter we described the main features of four narrative generators. MESSY, developed in the 1970s, shows how programming techniques and programming environments can be used as a framework for designing systems that generate narratives. SCÉALEXTRIC, a system that uses information recorded in knowledge repositories to generate stories, illustrates the capacity of computers to generate diverse narratives employing combinatorial techniques. GRIOT, which employs plots written by humans, shows the use of templates for interactive narrative. The chess-game narrator, which transforms the description of a chess game into a story, exemplifies programs that transform a set of data, not generated by the system, into a narrative. In summary, we studied systems that explore how general-purpose techniques commonly used in computer science for other purposes—i.e. programming techniques, combinatorial techniques, data analysis and rule-based techniques—can be employed for narrative generation. In Chapter 13 we will study other approaches.

13
Endless ways to use computers to study narrative generation (part 2)

13.1 CURVESHIP

CURVESHIP (Montfort 2011) is a computer program capable of telling the same story in different ways. That is, given a story, the system reorders the events to produce diverse narrative discourses. The program can also relate episodes from the perspective of different characters, narrate how a character describes the happenings of the tale to other character, and produce descriptions at different moments in time; that is, it takes a narrative specification about 'time of narrating' and generates text in the appropriate grammatical tense.

CURVESHIP is inspired by studies in narratology. The distinction between story (what is told about) and discourse (how the story is told) is at the core of the system. The work of authors like Genette (1980, 1988) and Prince (1982) shapes the design of this program. Let us illustrate this point with Genette's (1980) discussions about the order in which events are told. For him, events might be narrated in the order they occur (known as chronicle), in reverse chronological order (known as retrograde), interleaving events that happened in the past and in the now (known as zigzag), using flashbacks (known as analepsis), using flashforward (known as prolepsis), arranging the events based on some classification that is not chronological (known as syllepsis), and arranging the events in a way that it is impossible to determine the order they initially occurred (known as achrony). These concepts are the foundations of CURVESHIP's algorithms to generate diverse narrative discourses. In this way, one of the main contributions of CURVESHIP is to illustrate how knowledge generated in narratology can be represented in computer terms to develop automatized storytellers (for details see Montfort (2007a)).

The original version of the system, known as Curveship-py, makes it possible to design interactive fiction experiences with narrative variants. In this chapter we discuss Curveship-js, a shorter version that focuses on story variations. CURVESHIP requires that the user defines in the programming language JavaScript the elements that compose the narrative, namely, the locations where the actions take place, the actors participating in the text, things or objects relevant for the story, and the sequence of events performed by the characters. The system employs templates to generate its output; the variable part of these templates is designed to adapt to the diverse types of narration. We employ 'The Simulated Bank Robbery' (Montfort 2007b) to illustrate how the program works. This story describes a supposedly 'fake'

An Introduction to Narrative Generators. Rafael Pérez y Pérez and Mike Sharples, Oxford University Press.
© Rafael Pérez y Pérez and Mike Sharples (2023). DOI: 10.1093/oso/9780198876601.003.0013

bank robbery that results in a real death. The first step is to define the story's locations. For this example the bank includes a lobby, a vestibule, and a guard post; we also need a street so the thief can come into the building. Next, we define the actors; in this case we have the teller, the robber, and the guard. Now we define the things or objects relevant for the plot. This example includes a deposit slip, fake money, a bag, a mask, a fake gun, and a pistol.

CURVESHIP requires that the user provides data about characters, locations, and objects to generate its outputs. It is necessary to stipulate the text that the program uses to portray each character in the generated output, that is, its noun phrase. For this example, the thief is described as 'twitchy man', the teller as 'bank teller', and the guard as 'burly guard'. The user must indicate which pronouns (masculine or feminine are options, along with others) and which initial article is used for each character. In this exercise, the bank teller requires a feminine pronoun and her initial article is 'a'. So the first time that the 'bank teller' is mentioned in the story she is called 'a bank teller'; if this character is mentioned again, she is called 'the bank teller'. Similarly, the robber and guard use a masculine pronoun and 'a' as an initial article. Last, all characters must have an initial position. In this example, we situate the 'bank teller' in the vestibule of the bank, the 'twitchy man' on the street, and the 'burly guard' in his guard post. The information about the teller can be summarized as follows:

Character: Teller
Initial article: 'a'
Noun phrase: 'bank teller'
Pronoun: feminine
Location: in the vestibule

The system requires similar data for places and objects. Finally, the user introduces the sequence of actions that compose the story. They need to be typed in chronological order. Each action must indicate the doing (the verb phrase) and the characters and/or objects participating in it. CURVESHIP employs templates to translate actions into text. The user might also specify a template to indicate how to articulate a particular deed. The following lines show the first part of 'The Simulated Bank Robbery':

A bank teller reads a deposit slip.
A burly guard sleeps.
The bank teller rechecks the deposit slip.
A twitchy man puts on a Dora the Explorer mask.
The bank teller types.
She plays Solitaire a bit on her computer.
The twitchy man leaves the street.
The bank teller waves to him.
He threatens her using a gun-shaped object.
She laughs.
The burly guard wakes.
. . .

This story is told in the present tense and in the third person. The program exploits the use of pronouns, for example, 'She plays Solitaire a bit on her computer'; it also changes 'a bank teller' to 'the bank teller' in the third line (and similarly with the deposit slip) because she is no longer a 'new' element in the tale.

The next example shows CURVESHIP's output when the user asks the system to tell the same story in chronological order but, this time, the narrator is changed to the teller:

> I read a deposit slip.
> A burly guard sleeps.
> I recheck the deposit slip.
> A twitchy man puts on a Dora the Explorer mask.
> I type.
> I play Solitaire a bit on my computer.
> The twitchy man leaves the street.
> I wave to him.
> He threatens me using a gun-shaped object.
> I laugh.
> The burly guard wakes.
> . . .

The story is still narrated in present tense but, this time, all actions performed by the teller are described in first person. When necessary, the form of the verbs is automatically adjusted. Thus, instead of 'A bank teller reads a deposit slip', the system tells 'I read a deposit slip'; instead of 'The bank teller rechecks the deposit slip', the program writes 'I recheck the deposit slip'; and so on. In this way, we learn what happens in the bank from the perspective of the teller.

The last example shows the concluding actions of CURVESHIP's output when the user asks the system to tell the same story in retrograde order, with the guard narrating the story to the robber (the narratee).

> . . .
> I woke.
> The bank teller laughed.
> You threatened her using a gun-shaped object.
> She waved to you.
> You left the street.
> The bank teller played Solitaire a bit on her computer.
> She typed.
> You put on a Dora the Explorer mask.
> The bank teller rechecked a deposit slip.
> I slept.
> The bank teller read the deposit slip.
> The end.

Because this story is narrated backwards, the initial actions in the first example become the final incidents in this text. All verbs are in past tense (this is a designer's decision). All actions performed by the guard are described in first person, for example, 'I slept', 'I woke', while all actions performed by the thief are described in second person, such as 'You put on a Dora the Explorer mask', 'You threatened her using a gun-shaped object'. This story allows the reader to listen to the account that the guard tells to the thief about the failed attempt of robbery.

CURVESHIP explores how the same story can be told in different ways, a fundamental aspect in narratology that few computational models represent.

13.2 SLANT

SLANT (Montfort et al. 2013) is a storyteller that integrates some of the core features of three well-known systems studied in this book: MEXICA, CURVESHIP, and GRIOT. The antecedent of this project is MEXICA-nn (Montfort and Pérez y Pérez 2008), a pipeline system where MEXICA generated a plot and then CURVESHIP (back then called 'nn') determined how the narrative discourse would be arranged. By contrast, SLANT comprises five major subsystems that collaborate to generate stories in different genres that, optionally, incorporate figurative language based on a conceptual metaphor. The five components of the system are:

- MEXICA (the engagement–reflection module): its purpose is to progress a plot.
- Verso: this subsystem is informed by CURVESHIP's approach to genre and narrative theory. Its purpose is to choose the right genre for the story in progress and, when necessary, constrain the directions that a story might follow.
- Fig-S and Griot-Gen: these two subsystems are informed by GRIOT. Their purpose is to insert figurative language into the story.
- Curveship-Gen: this component is a specialized version of CURVESHIP. Its purpose is to realize diverse narrative discourses.

SLANT includes ten genres: confession, diary, dream, fragments, hangover, joke, letter, memento, memoir, play-by-play, prophecy, and the 'standard' story. The authors clarify that their use of the concept 'genre' is wide and that, in the context of this project, genre could also be understood as a style or mode.

The storyteller is organized as follows. Three components—MEXICA, Verso, and Fig-S—cyclically interact through a blackboard, that is, a computer data structure that provides a location where the three subsystems can write and read information to progress a narrative; none of them can modify or eliminate what other agents have written. In this way, each subsystem has the opportunity to contribute to the current task. MEXICA starts the loop, followed by Verso and Fig-S:

MEXICA → Verso → Fig-S → MEXICA → Verso → Fig-S . . .

For the experiments reported here, the authors set this loop to two cycles; this parameter can be modified. When the loops ends and the specifications for the new tale are completed, Griot-Gen realizes the figurative language and finally Curveship-Gen produces the final output in English. These are the steps that SLANT follows to produce a new narrative:

(1) MEXICA starts the process by writing on the blackboard a partial plot generated during one cycle of engagement and reflection.

(2) Verso reads the blackboard and decides the type of narration to be used in the story; in the current implementation the options are personal (first person) or impersonal (third person). The system looks at the work in progress for a character present in the story locations where most of the actions take place and who either is involved in a few of those events (more like a witness) or is the object of lots of those deeds. If such a character exists, the type of narration is set to personal; otherwise, it is set to impersonal (this process requires that the designer define a number of incidents that represent 'few' and 'lots' of actions). Verso might also set some constraints to influence the way MEXICA will progress the story. For instance, if the type of narration is set to personal, then the program requires that the narrator never dies. These pieces of information are written on the blackboard.

(3) Fig-S reads the blackboard. At this point Fig-S does nothing because in the current version it only works at step (6), that is, when the plot has been finalized. But conceptually, each subsystem can participate once in every cycle.

(4) MEXICA reads the blackboard and, taking into consideration Verso's requests, further elaborates the plot until it is completed. Then the final plot is written on the blackboard.

(5) Verso reads the blackboard and searches the story for specific features, like actions with particular verbs, patterns of actions, or sets of characters, which are used to choose the best genre for the tale in progress. For instance, the 'confession' genre is picked when most of the action in the narrative can be considered as sins (robbing, killing, etc.) based on a list of these. Then Verso selects the narrator, the narratee, time of narrating, rhetorical style, and phrases to begin and/or end the story that correspond to the chosen genre. This information is written on the blackboard.

(6) Fig-S reads the blackboard and updates it to include metaphorical content. This subsystem employs a collection of predefined domains and concepts. For instance, the domain 'fire', which is useful to represent metaphors like 'Love is fire', includes concepts like 'burn' and 'incendiary'. In this way, Fig-S analyses each action in the story to determine which of them can be linked to any of the existing domains and then be replaced by metaphorical adaptations of the same deed. The subsystem can perform this process in two ways: either it can find a single domain that best suits the largest number of actions or it can find an appropriate domain for each action in the plot. The first option produces a more coherent output while the second method results in a greater diversity.

At this point, the loops ends.

(7) Griot-Gen reads the blackboard and produces templates for natural language representations for figuratively versions of actions.
(8) Curveship-Gen processes the information produced by MEXICA, Verso, and Griot-Gen to produce the final text based on templates.

The following shows a story generated by SLANT in the 'confession' genre with the figurative module off. This represents a story as if it were told to a priest at confession. Verso chooses as a narrator the character that performed that last sinful action. The story is narrated in past tense and employs a 'hesitant' style that incorporates expressions like 'um' and 'er' to indicate that the narrator is anxious. All stories developed in the 'confession' genre always start with 'Forgive me, Father, for I have sinned. It has been a month since my last confession' and end with the conclusion 'Ten Hail Marys? Thank you, Father.'

> Forgive me, Father, for I have sinned. It has been a month since my last confession. An enemy slid. The enemy fell. The enemy injured himself. I located a curative plant. I cured the enemy with the curative plant. The tlatoani kidnapped me. The enemy sought the tlatoani. The enemy travelled. The enemy, um, looked. The enemy found the tlatoani. The enemy observed, uh, the tlatoani. The enemy drew a weapon. The enemy attacked the tlatoani. The enemy killed the tlatoani with a dagger. The enemy rescued me. The enemy entranced, uh, me. I became jealous of the enemy. I killed the enemy with the dagger. I killed myself, uh, with the dagger. Ten Hail Marys? Thank you, Father.

This example is useful for analysing some of SLANT's features. The narrator confesses to the priest the murder of the enemy and then he states that he killed himself: 'I killed the enemy with the dagger. I killed myself, uh, with the dagger.' This scene is the result of some fortuitous technical incidents. When this story was generated Verso was not able to instruct MEXICA to avoid certain actions, like killing the narrator. This story made evident to the authors the necessity of having a richer communication between Verso and MEXICA, where the former could influence the process performed by the latter, a feature that was implemented later. That is why the narrator dies at the end. Some readers might find this situation incongruent. However, others might have a different interpretation. Maybe the narrator is in purgatory and going to confession there.

Montfort and his colleagues point out another possible reading of this tale. When the Spanish conquerors arrived in México, they imposed Catholicism on the natives. Thus, the authors suggest that for MEXICA to generate plots narrated as Catholic confessions might be seen as a provocation about the colonial history of México that contributed to the tragic destruction of the Mexica civilization. The authors make clear that the system does not represent historical knowledge such as that mentioned here and, therefore, cannot produce or suggest standpoints about historical facts. Rather, this type of thoughts are the result of readers' own experiences. However, it must be pointed out that it was the decision of the SLANT designers to include a 'confessional' genre in the context of MEXICA. In any case, this does demonstrate how story generators can provoke discussion about genre and authorship.

In this way, different readers might have different interpretations of the same text even when SLANT cannot be credited for this. Montfort and his colleagues suggest that this type of narratives show how the collaboration between MEXICA, Verso, and Curveship-Gen in SLANT might generate cultural resonance products that are surprising and significant by being thought-provoking.

The following shows a story with figuration. The genre chosen is 'play-by-play' based on sports commentary. Some of the keywords that Verso searches for in the plot to select this genre are 'bleed', 'cure', 'close on', 'draw', 'escape', 'fall', 'fight', 'panic', and 'wound'. The prose style is 'excited'. In this tale Fig-S attempts to find an appropriate metaphor for each action in the plot. As a result, this narrative includes a diversity of less consistent metaphors.

> This is Ehecatl, live from the scene. The cold-wind Eagle knight is despising the icy jaguar knight! The cold-wind jaguar knight is despising the chilling eagle knight! Yes, an eagle knight is fighting a jaguar knight! Look at this, the eagle knight is drawing a weapon! Look at this, the eagle knight is closing on the jaguar knight! The gardener eagle knight is wounding the weed jaguar knight! And now, the jaguar knight is bleeding! Yes, the consumed eagle-knight is panicking! And, eagle knight is hiding! Holy—the snowflake slave is despising the chilling jaguar knight! The freezing-wind jaguar knight is despising the cold slave! And, yes, the cold-wind slave is detesting the chilling jaguar knight! A slave is curing the jaguar knight! And, the slave is returning to the city! And, the jaguar knight is suffering! The frozen jaguar knight is dying! Back to you!

The story is narrated in third person and present tense. The opening and closing sentences are 'This is Ehecatl, live from the scene' and 'Back to you!', respectively. Expressions like 'Look at this' or 'Holy' are typical of the 'excited' prose style. Finally, 'the weed jaguar knight!', 'the snowflake slave', and 'the chilling jaguar knight!' exemplify the type of figurative language generated by the system.

This project has faced several difficulties. One of the main challenges has been to manage the differences in the design of fundamental aspects of these programs. Here are some examples. Characters and locations are represented in MEXICA and CURVESHIP; however, the latter also characterizes 'things', a concept that does not exist in the plot generator. MEXICA was adjusted for use in SLANT to produce representations of things mentioned in actions. In CURVESHIP, all actions correspond to a sentence with a single verb phrase. By contrast, MEXICA might employ complex actions like 'Character-A tried to intimidate Character-B with a weapon but instead he wounded himself' (this intricate situation is represented in MEXICA as the single action 'Character-A Faked_Stab_Instead_Hurt_Himself Character-B'). The solution was to develop a mapping between MEXICA actions and SLANT actions. Similarly, CURVESHIP's templates were designed for narrative variation, so they had to be adapted in SLANT to work with figurative text. Some other issues have not been fully solved yet. MEXICA implements disnarration (Prince 1988); that is, the program not only describes what happened in the tale but also what did not occur, for example, 'Character-A did not cure Character-B'. CURVESHIP, and therefore Verso and Curveship-Gen, are not intended for processing disnarration. This issue will be solved in further versions.

MEXICA, CURVESHIP, and GRIOT were designed independently to study dissimilar aspects of automatic narrative generation. SLANT illustrates how three systems that pursue different aims can collaborate to produce narratives that none of them could have developed individually.

13.3 DAYDREAMER

DAYDREAMER (Mueller 1987, 1990) is a system that models the process of daydreaming. The program has two operations modes that we refer to as Real-world and Daydreaming. The Real-world mode represents activities that a character, known as the daydreamer, performs in a hypothetical real world, such as going to a theatre or meeting people in a restaurant. The purpose of performing those actions is to satisfy a set of predefined necessities that the daydreamer has, such as establishing romantic relationships, having entertainment, eating, and earning money. Depending on whether these goals are achieved, the character displays a positive or negative emotional reaction. These emotional states are a fundamental for the system because they trigger the process of daydreaming; that is, they activate the Daydreaming operation mode. During the Daydreaming mode, the system generates what Mueller describes as a stream of thought that recollects past incidents, pictures alternative directions for those past events, and fantasizes future experiences. The program is constantly switching between these two operation modes.

The design of the model is based on a profound study of the literature of the time about daydreaming in humans and about problem-solving in AI. Mueller examined several daydreaming reports, which were produced employing diverse techniques like retrospective reports, think-aloud protocols, thought sampling, and event recording, to identify relevant features of the process. The work is based on the premise that daydreaming is relevant for learning from imagined incidents, for developing creative problem-solving skills, and for the useful interaction with emotions to overcome failures. In this way, the process of daydreaming enables the ongoing revision of previous solutions, explores what might seem as unrealistic possibilities to solve a problem, and supports the accidental recognition of relationships among different problems to be solved (serendipity). Furthermore, daydreaming is useful for reducing or eliminating negative emotions resulting from past failures; in other words, it helps people to feel better. These features are the foundational ideas of DAYDREAMER.

DAYDREAMER attempts to generate similar outputs to those think-aloud protocols that Mueller analysed. The system is based on the problem-solving and planning techniques that we studied in Chapters 3 to 7. Thus, it attempts to achieve a top-level goal by breaking it into subgoals, which in turn might be broken further into other subgoals, and so on. The model employs two types of goals known as personal goals and daydreaming goals. Personal goals characterize everyday necessities that need to be fulfilled; the program represents thirteen personal goals, known as self-esteem, social-esteem, lovers, friends, employment, food, sex, love-giving, love-receiving, companionship, entertainment, money, and possessions. Daydreaming goals characterize mechanisms of dealing with the emotional reactions that result from real-world events; the program represents seven daydreaming goals, known as rationalization, roving, revenge, reversal, recovery, rehearsal, and repercussions. Thus, the purpose of daydreaming goals is to provide the system with structures and plans, equivalent to the themes we studied in Chapter 6, to generate narratives that are useful for dealing with negative emotions produced by real events. These are the descriptions of each daydreaming goal:

- *Rationalization.* The daydreamer produces a narrative where the disappointment produced by a personal goal failure is reduced by creating a context for consolation. For instance, the system develops a situation where the illusory success of a goal that in reality has failed results in an imagined goal failure; or a scenario where the failure of a goal is useful for achieving a second real or imaginary goal.
- *Roving.* The daydreamer generates a narrative where the focus of her thought changes from a personal goal failure towards a positive past episode or an imagined scenario. As a result, the negative emotional state produced by the failure is reduced.
- *Revenge.* The daydreamer generates a narrative where she takes revenge on a person that thwarts one of her personal goals.
- *Reversal.* The daydreamer generates a narrative where she avoids a real past personal goal failure or where she prevents an imagined goal failure in the future.
- *Recovery.* The daydreamer generates a narrative where a previously failed personal goal is achieved.
- *Rehearsal.* The daydreamer pictures a narrative where she achieves a personal goal.
- *Repercussions.* The daydreamer explores the possible outcomes of a future situation.

Rationalization, roving, and revenge produce situations that activate positive emotions; they are used to reduce the intensity of the currently active negative emotions. Reversal, recovery, rehearsal, and repercussions allow the daydreamer to learn from real and imagined setbacks, as we explain in the following.

The system characterizes emotions like pleasure, amusement, satiation, displeasure, regret, pride, shame, embarrassment, and heartbreak. They are represented as structures that include attributes like a sign (positive or negative), a strength (a number specifying the magnitude of the emotion), the person towards whom the emotion is directed, and the goal from which the emotion resulted as a reaction. Emotions are activated in response to the achievement or failure of goals in real-world situations. For instance, when the daydreamer eats, satiation is activated; when she fails to establish a romantic relationship, heartbreak is activated. So the emotional state of the system comprises all active emotions.

In this way, emotions trigger daydreaming goals. Daydreaming might modify emotions' attributes or start new emotions, which in turn might trigger new daydreaming goals, and so on. Although all Daydreaming goals are triggered by emotions, the conditions that start each of them might change from goal to goal. For instance, negative emotions like embarrassment, rejection, and anger are part of the preconditions that motivate rationalization, roving, reversal, and recovery; while, to be triggered, revenge requires an active negative emotion directed towards another person.

The program employs diverse strategies for achieving goals that are encoded as inference and planning rules. Inference rules describe the consequences of a given situation while planning rules indicates how to decompose a goal into subgoals. The following shows an example:

Inference rule
If one likes something that one believes another person likes,
then one will like that person.

Planning rule
If one has a goal to like someone,
then the subgoal may be created for one to like something that one believes the
other person likes.

In some cases, the system includes several alternatives for breaking down a particular goal. In this way, the set of predefined rules allows the system to generate diverse and coherent daydreams.

The generation process in DAYDREAMER is complex. It involves a constant interplay between the real-world mode, the daydreaming mode, and some data provided by the user in real time (in this way, the user can influence part of the real-world experience). Here we present a very general description of how the system works. For this example, the first step is for the user to define the initial context of the real-world (i.e. the initial value of some variables). In this case, the daydreamer is a woman, she has a job, she is not involved in an amorous relationship, she is romantically interested in the actor Harrison Ford, she is currently at home, she knows where the Nuart Theatre is located, and so on. This definition also establishes those necessities (or goals) that the character needs to fulfil.

Next, the system simulates a situation in the real world that motivates the process of daydreaming. It works as follows. When DAYDREAMER starts, it triggers two goals, one to establish a romantic relationship with someone (lovers) and another to satisfy a need for recreation (entertainment). In this way, the program might have two or more active goals at the same time; an internal mechanism controls in which of those active goals the system must focus at any given time. To satisfy the entertainment goal, she decides to go see a film at the Nuart Theater where she accidentally runs into Harrison Ford. Because she always has been interested in him, and she is looking for a romantic relationship, the daydreamer decides to invite Harrison to go for a dinner; unfortunately for her, he turns her down. This failed experience triggers daydreaming goals like revenge, rationalization, roving, recovery, and reversal. The following illustrates the outcomes of these daydreaming processes (they were all published in Mueller 1990):

Revenge

I study to be an actor. I am a movie star even more famous than he is. I feel pleased. He is interested in me. He breaks up with his girlfriend. He wants to be going out with me. He calls me up. I turn him down. I get even with him. I feel pleased.

Rationalization

What if I were going out with him? He would need work. I remember the time he had a job with Paramount pictures in Cairo. He would go to Cairo. Our relationship would be in trouble. I would go to Cairo. I would lose my job at May Company. I feel relieved.

Roving

I remember the time Steve told me he thought I was wonderful at Gulliver's. I feel pleased. I remember the time I had a job in the Marina. I remember the time Steve and I bought sunglasses in Venice Beach.

Recovery

I have to ask him out. I have to call him. I have to know his telephone number. He has to tell me his telephone number. I have to know where he lives. Suppose he tells someone else his telephone number. What do you know! This person has to tell me his telephone number. He has to tell this person his telephone number. He has to want to be going out with this person. He has to believe that this person is attractive. He believes that Karen is attractive. He is interested in her. He breaks up with his girlfriend. He wants to be going out with her. He wants her to know his telephone number. She tells him she would like to know his telephone number. He tells her his telephone number. I have to tell her that I want to know his telephone number. I call her. I tell her that I want to know his telephone number. She tells me his telephone number. I call him. I ask him out.

Reversal

I feel embarrassed. What if I had put on nicer clothes? I would have gone to the Nuart Theatre. I would have asked him out. He would have accepted. I feel regretful.

Each time DAYDREAMER generates a new episode, the system records all the steps completed to accomplish the task. This information is used in different ways.

First, when a new daydream is in progress, the system first attempts to use its past experiences (similar resolved goals) as a guide to achieve the current goal; this process is called analogical planning (cf. case-based reasoning studied in Chapter 7). Second, a given situation might be modified to find new ways to solve a problem. For instance, the action 'Harrison Ford tells me his telephone number' might be transformed into 'someone else tells me Harrison Ford's telephone number', or into 'Harrison Ford tells some else his telephone number' (like the in previous recovery daydreaming example). This process is called mutation (cf. TRAMS studied in Chapter 7). Third, while developing new scenarios, the system might retrieve an episode that is useful for accomplishing a different goal to the one the system is currently focused on achieving. The program is constantly checking for this possibility to occur. This process is called serendipity. Thus, DAYDREAMER cannot learn daydreaming goals; all of them are hard-wired. However, the system is capable of learning novel ways of solving the existing daydreaming goals through analogical planning, mutation, and serendipity. Finally, the system employs a text generator, based on templates, to transform internal data structures into written English.

13.4 There are many more systems . . .

In this chapter we have illustrated how the knowledge generated in other fields shapes the design of programs for narrative generation. CURVESHIP illustrates how studies in narratology can be represented in computer terms to develop automatized storytellers. We have described how the design of DAYDREAMER is based on research into daydreaming in humans. We have also discussed SLANT, a system that that integrates some of the core features of MEXICA, CURVESHIP, and GRIOT (all these systems are studied in this book).

Thus, in Chapters 12 and 13 we have described seven systems that illustrate a diversity of topics and methods for computer models of narrative generation. We could write a whole book describing the general characteristics of other interesting systems. For instance, a classic storyteller is UNIVERSE (Lebowitz 1985), which develops stories in a soap-opera style that, like dramas written for television, seem never to end; FAÇADE (Mateas 2002) is an interactive drama system that explores how the interaction of three characters (two controlled by the program and one by the user) and the use of a drama-managed plot module give rise to a story; FABULIST (Riedl 2004) studies characters' believability, that is, how to generate narratives where characters show intentions driven by their beliefs, desires, and traits; RUI (Ontañón and Zhu 2010) explores the use of analogy for story generation; PREVOYANT (Bae and Young 2014) studies the use of flashback and foreshadowing to produce surprise in a planning-based narrative generator; ISLANDERS (Ryan 2018) is a two-step program: first, the system plays games in a story-world and then the program transforms the interesting events that occurred during the game into stories; and so on. For a review of computer systems for narrative generation see Gervás (2009), Kybartas and Bidarra (2017), Hou et al. (2019), Herrera-González et al. (2020), Alhussain and Azmi (2021).

In the final chapter, Chapter 14, we discuss the general features of narrative generators and the consequences of these types of system for society.

14
The story we always wanted to tell you

14.1 A global view of the field

The models considered in this book illustrate how computers can be used to characterize different perspectives and ways of studying narratives. We have studied five techniques in detail in this book. Here is a summary:

(a) *Templates*. Templates require that the designer of the system defines the following elements: a set of fixed texts that contain fields to be filled with data, ways to obtain such data, and mechanisms to combine templates in a coherent way. The structure and coherence of the narrative is established by the predefined text. In this approach, the human designer makes all the decisions about the content and organization of the narrative while the computer fills in the fields with relevant data.

(b) *Problem-solving*. In problem-solving, narratives are built around characters' goals and plans to meet those goals. Plans assist the coherence of the plot. The complications that might arise when characters attempt to achieve their goals allow the system to represent an essential element of any story: its conflict. The characterization of notions like the need for love is useful for developing interesting scenarios. From a technical point of view, problem-solving is seen as a search to find suitable combinations of data structures that represent sequences of actions. These algorithms exploit the capacity of computers to evaluate rapidly an enormous number of possible data-structure permutations. The story ends when all characters' goals in the system have been fulfilled.

(c) *Planning*. The planning approach complements problem-solving by introducing a new element: the author's perspective. In addition to representing the characters and their desires, this approach also characterizes an author's goals, plans that meet those goals, and the story's thematic structures. There is a clear distinction between the goals of the author and those of the characters. Compared to the previous technique, the technical complexity of these systems increases because now it is necessary to use more elaborated data structures, as well as to build mechanisms to handle when the system focuses on solving the author's goals, when it focuses on solving the characters' goals, and when it determines to put one goal on hold to meet another. As in the previous case, planning represents automatic narrative generation as a search process that ends when all the goals in the system have been fulfilled. The influence of this methodology continues nowadays.

(d) *Engagement and reflection*. This technique produces narratives that are the result of a cycle between two states of creative writing known as engagement and reflection.

An Introduction to Narrative Generators. Rafael Pérez y Pérez and Mike Sharples, Oxford University Press.
© Rafael Pérez y Pérez and Mike Sharples (2023). DOI: 10.1093/oso/9780198876601.003.0014

The representation of goals, plans, and story-structures is avoided. During engagement, the constraints produced by the emotional relationships and conflicts between characters drive development of the plot. During reflection, the material generated so far is evaluated and improved. This type of system represents an author in a continuous mental cycle between engagement and reflection.

(e) *Statistical methods.* Statistical methods, such as Markov chains and deep neural networks (DNNs), focus on obtaining statistical relationships between the elements that make up their database. Large DNNs are capable of extracting millions of structures from their databases, and of generating probabilistic networks to combine them in a useful way. The representations of explicit abstract concepts such as characters' goals, plans, or tensions have no place in this technique, although they might generate narratives 'as if' the characters had goals, plans, and tensions. We can picture DNNs as systems that represent an author capable of representing abstract data structures that advance sequences of words that form sentences and then narratives.

Although this classification is useful as a general framework for discussion, we should be aware that some systems may include features from different categories. For example, although DAYDREAMER (see Chapter 13) is a planning-based system, it only represents the author's goals. That is, the author is also the core character in all its stories (other characters are not represented as data structures in the system). In DAYDREAMER, the distinction between author and characters, typical of planning, is blurred. Other systems, like those based on rules and permutation heuristics (see Chapter 12), might mix elements of diverse categories. We should also consider that there are narrative generation methods not discussed in this book. Thus, this classification should not be used in a rigid way.

14.1.1 Audience versus theory

Jordan and Russell write that 'There are two complementary views of artificial intelligence (AI): one as an engineering discipline concerned with the creation of intelligent machines, the other as an empirical science concerned with the computational modelling of human intelligence' (Jordan and Russell 1999, lxxiii). These authors refer to the former view as modern AI and to the latter view as describing an important part of modern cognitive science. Following these authors, we have defined two similar categories for automatic narrative generation: audience approach and theory approach. These are their features:

- *Audience approach.* Emphasis is on the construction of products that are appealing to an audience or useful for performing a task such as assisting a human author. This type of narrative generator is mainly based on mathematical models and engineering methods. Researchers have figured out how to engineer those tools to develop computer programs that produce narratives.
- *Theory approach.* This type of system is designed according to theories developed in the humanities and the social sciences as frameworks to drive its design.

Studies from narratology, human daydreaming, cognitive science, and so on are the basis for developing computer models. These models are then tested as running programs that work as prototypes. Thus, the elements of the theory have an explicit correlation with the components that compose the prototype. Emphasis is on contributing to the understanding of narrative generation by providing potential explanations about how this process might work.

Language models, like those we studied in Chapter 10, illustrate systems with a tendency towards the audience approach. Other programs are clearly located at some midpoint between these two categories. In Chapter 12 we employed GRIOT to exemplify the use of templates. However, GRIOT also includes a module that represents the cognitive theory of conceptual blending. SLANT employs an engineering architecture known as a blackboard to communicate its elements; however, its main modules are based on MEXICA, CURVESHIP, and GRIOT, systems associated with a theory approach.

MEXICA, based on the engagement–reflection cognitive account of creative writing, represents systems with a tendency towards the theory approach. Other examples of the theory approach are TALE-SPIN and MINSTREL. Meehan explains that TALE-SPIN, a system based on problem-solving techniques, focuses on studying what humans require knowing to make up stories; to achieve this goal, his system integrates diverse theories and sources of information into one program (Meehan 1976, 2). Similarly, Turner claims that 'MINSTREL is based on a model of the author as a *problem solver*' (his emphasis; Turner 1994, 10). He also states that MINSTREL follows the guidelines that Weisberg (1986) gives for a model of creativity, includes features of the Wallace model (1926), and combines all these ideas with other thoughts not included in any of these theories (Turner 1993, 537). That is, this program combines a model of writing as problem-solving with a model of creativity.

Thus, the audience versus theory classification is useful for analysing whether a program has been designed to contribute to the understanding of the writing process, or whether it has been designed to produce coherent texts, without much interest in shedding light on how we write.

14.1.2 Endless ways to reflect on automatic narrative generation

There are still further ways to classify narrative generators based on relevant features. For instance:

- *Realization of the story*. In narratology, there is a difference between the events that occurred in a story and how those events are described to the reader. Let us explore this distinction from the perspective of automated narrators. Most of the systems we have studied in this book first generate a complete plot, usually represented inside the system as sets of data structures, and then employ templates to 'translate' those structures into the text presented to the reader. Examples are TALE-SPIN, MINSTREL, MEXICA, and DAYDREAMER. In a way, the plot generation process is detached from the realization of the story. Thus, it should be possible to add to these programs new modules that generate different types of output without modifying the process that generates a story or changing the content of the knowledge base, for example, texts in different languages, visual narratives, and multimedia. By contrast, in most of the language models that we have analysed, the plot develops as the words and sentences that conform the output are assembled together. As we discussed in Section 10.5, because in these automated narrators the words shape the plot during the creation process, words and plot are strongly related. Systems based exclusively in templates (like those in Chapter 2) employ plots provided by the designer of the program. None of these methods alone will produce a full model of narrative creativity. The plot systems rely on human-crafted sentences to present the plot as a narrative. The DNN systems have no explicit representation of plot and so cannot manipulate story-structure. The template systems have pre-prepared story-structures. To design a full narrative generator will require combining at least two of these, possibly all three.

- *Coherence.* One of the most critical aspects for a narrative generator is the production of coherent outputs. The use of predefined story-structures, predefined plans, or both is the most common method used to achieve this goal. So far, only a few systems, such as those based on DNNs or the engagement–reflection approach, build in real time the structure of the narrative they produce.
- *Creative storytelling.* Only three of the systems we studied have an explicit model of the creative process: DAYDREAMER represents creativity through serendipity and analogical reasoning, MINSTREL employs TRAMS, and MEXICA uses the engagement–reflection cognitive account of creative writing. Self-evaluation is an essential part of creative narrative generation. MINSTREL is capable of evaluating the novelty of the story it is producing while MEXICA evaluates the novelty, interestingness, and coherence of its outputs (you can find a description of the core features of creative computer-based storytellers in Pérez y Pérez and Sharples (2004) and Pérez y Pérez (2015b).
- *Cultural representations.* MEXICA and GRIOT have been used to implement cultural forms of narrative that are not often privileged in computer science, in this case, descriptions of some elements of an ancient culture from México and the oral traditions of narrative from the African diaspora (Montfort et al. 2013).

The categories introduced in this chapter illustrate the extraordinary flexibility of computers for capturing different or even discordant visions about narrative generation. However, they also make evident the necessity of developing mechanisms that make it possible to classify narrative generators in an objective way; beside the claims that authors might make about their own programs, a common framework that allows independent experts to analyse and compare these systems on the same basis is needed. The community still needs to work on these methods.

14.1.3 Final remarks about the systems we have studied

Table 14.1 shows a summary of the main features of the programs we have studied in this book. The table indicates the year each program was first reported.

Recently, programs based on DNNs have come to dominate the field. As we have studied, all approaches have been important for advancing this area and all of them have limitations. The study of story generation will benefit if researchers follow a more holistic approach where all types of system are developed and where different ways of combining the best of each approach are explored. SLANT is the only project we are aware of that integrates the core features of multiple programs that pursue different narrative goals. Nevertheless, there are some advances in this direction.

Table 14.1 Description of the core features of the automatic narrative generators studied in this book

Narrative generator	Description
MESSY (1976)	A behavioural simulation programming language and a simulation system that executes the instructions specified through the language.
TALE-SPIN (1976)	Automatic narrative generation based on problem-solving techniques.
DAYDREAMER (1987)	Computational representation of creative daydreaming through planning and the use of emotions to condition the development of the narrative.
MINSTREL (1993)	Representation of a model of writing as planning and problem-solving combined with a model of creativity. The core element of the model of creativity is a set of heuristics known as TRAMS.
MEXICA (1999)	Computational representation of the creative process through the engagement–reflection account of writing and the use of emotions as a way of generating a sequence of story-actions.
GRIOT (2005)	Use of templates and a computational representation of figurative language based on the theory of conceptual blending.
CURVESHIP (2011)	Computational representation of narratology theories.
CHESS-NARRATOR (2012)	Use of rules to transform the description of a chess game into a story. It illustrates programs that convert a set of data not generated by the system into a narrative.
SLANT (2013)	Integration of the core features of three independent systems: MEXICA, CURVESHIP, and GRIOT.
SCÉALEXTRIC (2017)	A knowledge-based system for narrative generation that exploits the power of combination.
A noise such as a man might make (2018)	Use of Markov chains to generate a novel.
DNNs (2017–20)	Use of probabilistic networks to generate abstract data structures.

Our review of some illustrative systems in Chapter 10 shows how some researchers are attempting to incorporate problem-solving and story-structure concepts into the development of DNNs. Interdisciplinary approaches open marvellous unexplored opportunities.

There are techniques not studied in this book that have been used to develop narrative generators; for example, story grammars, genetic algorithms, natural language processing, and first-order logic. We hope that in the future we will have the opportunity to discuss all these valuable ideas. In Section 14.2, we look at how narrative generators influence society.

14.2 Towards a humanistic automatic-narrative generation

The advancement of intelligent-systems technologies is producing novel social, cultural, political, and economic contexts. A critical society must promote studies about the impact of AI in our everyday life, for example, studies on how people's beliefs about the capacities of automatic narrators are built and how those beliefs influence their expectations about the scopes and limitations of these systems. Similarly, a critical AI society must support and encourage the development of projects that benefit humanity and whose products are available to anyone. In the following we elaborate these ideas.

14.2.1 The effect of the mystery

Computer models provide scientists with the possibility of studying in detail their operation. However, although neural networks employ mathematical models in their implementation, experts cannot give detailed explanations that account for the processes that these systems generate internally to produce their narratives. That is, we do not fully understand the results of the processes of generalization, pattern detection, internal representation of information, among others. Developing methods that enable this limitation to be overcome is one of the main challenges in the field. We hope that, sooner or later, computer scientists will succeed in this endeavour; however, it is an open question as to how far researchers can 'decode' DNNs—and whether doing so will really help build more effective narrative generators. In any case, at the moment, we lack this knowledge. Despite this, it is common to find blogs in which authors attribute cognitive properties to DNNs that are impossible to verify (as we discussed in Chapter 1, this phenomenon is not exclusive of DNN). Similar situations occur in some academic environments. Why do some people have this penchant for making unsupported claims about the operation of a computer program? We believe that sometimes scientists make unsupported claims in an effort to achieve clarity and neatness (to 'solve the mystery'). We hope that one day psychologists will give us a convincing explanation of this phenomenon. Now we can only speculate about it. Let's do it.

As mentioned in Chapter 1, for years computers have generated texts that are difficult to distinguish from those written by humans. Every day news agencies employ template-based systems to publish hundreds of news and sports stories that many people believe are written by reporters. However, these systems do not capture the public imagination nor provoke controversy among scientists with the intensity that DNN-based systems do. There are several reasons for this, such as the ability of neural networks to generate convincing texts on a huge variety of topics, their colossal size, and the marketing plan behind the organizations that own the commercial rights to some of these systems. But there is a point about which little is said that is worth reflecting on. We have created AI systems that generate narratives in a way we do not fully comprehend. How much does the lack of understanding of the operation of these systems, which we refer to as *the effect of the mystery*, influence the perception that people and some specialists have about the capabilities of these systems? We suspect that the effect of the mystery contributes to some people making unsupported claims about computer programs. But, regardless of whether we are right or not, what matters here is to understand that the automatic generation of narratives is creating new social and technical contexts that need to be studied.

14.2.2 The story people want to hear

In June 2021, *The Atlantic* published an article about UK Prime Minister Boris Johnson.[1] This text argues that, for Johnson, politics is about providing people a story they can believe in. The author, Tom McTague, quotes Johnson as saying that 'People live by narrative . . . Human beings are creatures of the imagination.' Then McTague reflects on how Johnson is brilliant at understanding the stories voters want to hear. In the case of Brexit, it was a story that offered hope, agency, optimism, and pride. Can computers write the stories people want to hear? Can computers produce texts that offer hope, agency, optimism, and pride to human readers in challenging circumstances? What would be the readers' reactions to those narratives? We believe that the answer to the first two questions is yes, they can. We do not have yet an answer for the last question. So far, we know that intelligent systems are able to prompt social behaviours and trigger people's imagination. Therefore, it seems reasonable to think that these types of story might resonate, particularly among those under tough conditions. Another important point that needs to be discussed is attributing responsibility for such stories. There needs to be a debate about legal and ethical responsibility for dangerous, pernicious stories generated by computer. A critical AI society needs to promote research that analyses the pros and cons of using automatic narrators in situations like the one we have just described.

---

[1] Tom McTague, The minister of Chaos. *The Atlantic* (July/August 2021). Available at https://www. theatlantic.com/magazine/archive/2021/07/boris-johnson-minister-of-chaos/619010/

14.2.3 Intercultural diversity through storytelling

The project A Million Stories[2] is an intercultural storytelling project that gathers more than 600 stories told by refugees who have fled to the European Union in recent years. Those are life stories where refugees narrate the type of life they had lived before, the type of life they are living now, their fears, hopes, and dreams. A core goal of this project is to promote understanding and respect for intercultural diversity through storytelling. In this way, it is hoped to enhance the connection between the host population and refugees. Can automatic narrative generators contribute to this type of project? We claim that the answer is yes. In the following, we suggest three projects that have the potential to support initiatives like A Million Stories and, at the same time, provide research contexts to study the effects that computer-generated narratives have in readers.

(a) *An automatic narrative assistant.* Picture an automatic narrative generator like GPT-3 (see Chapter 10) that, using its enormous dataset, supports refugees to find appropriate ways for expressing their experiences in text. The user types some central ideas about her life and experiences and then the system employs those notions to progress a narrative. If the user is not satisfied with the results, she can start all over again; or she can modify the generated text and then feed it back to the system to try once more. This tool might be useful for people with difficulties speaking the host country language, and for individuals who find it hard to express their emotions and life experiences in writing. This proposal might help to answer research questions like these:

> Is the ability of GPT-3 to generate narratives from a set of significant words, or phrases, or a plot, enough to help refugees communicate their experiences?
> What are the refugees' reactions to collaborating with an AI system for such personal, and presumably sometimes painful, life experiences?
> Can collaborative narratives motivate refugees to share their experiences? Or, on the other hand, are automatic generated texts rejected by refugees?

And so on.

(b) *Refugees' life stories as inspiration for automatic narrative generators.* The purpose is to use techniques like planning or case-based reasoning to develop an automatic narrative generator that uses the refugees' narratives as sources of information to create its knowledge base. For instance, imagine a system like DAYDREAMER (see Chapter 13) that daydreams about refugees' fears and expectations. Employing the routines described in Chapter 13, this system generates narratives about refugees' lives where failures of past personal goals are avoided, or failed goals are achieved, or narratives that explore the possible outcomes of a future situation in the host country, and so on. This proposal might be useful for answering research questions like these:

---

[2] A Million Stories: Refugee lives. Available at http://refugeelives.eu, accessed 22/November 2022.

> Can automatic narrative generators represent human experiences such as those experienced by refugees?
>
> Can planning-based narrative generators represent (some of) the goals and plans spoken by refugees in their narratives?
>
> Can AI systems like DAYDREAMER generate narratives that embody the fears, hopes, and dreams articulated by refugees?
>
> Are the narratives generated by a system like DAYDREAMER, which represent refugees' fears and expectations, somehow be relevant to refugees? That is, do refugees identify with the experiences narrated by a program like that?

And so on.

(c) *Systems for the generation of collaborative narratives.* In many host countries there are residents who express concerns about receiving refugees and migrants. Because one of the goals of A Million Stories is to enhance the connection between the host population and refugees, we believe it is worthwhile to investigate whether AI narrative generators might be able to help in this respect. Let us elaborate this idea. In Chapter 11 we described MEXICA. We explained that this system employs human-written narratives, known as the previous stories, to create its knowledge base. Also, we briefly introduced the MEXICA-impro project. Its goal is to design two different MEXICAs programs, referred to as MEXICA-1 and MEXICA-2, each with its own knowledge base, that is, with its own set of previous stories, to work as a team to generate a collaborative narrative. Thus, in order to jointly advance a story's plot, it is necessary for the computational agents to find ways of reconciling the differences in their knowledge bases. In this way, MEXICA-impro generates texts that only can emerge as a result of the collaboration of both programs. Taking as a starting point the current state of MEXICA-impro, it is possible to advance these ideas as follows:

> Gather a group of previous stories written by refugees.
> Gather a second group of previous stories written by concerned host residents.
> Feed MEXICA-1 with the first set of narratives and MEXICA-2 with the second.
> Generate collaborative stories between MEXICA-1 and MEXICA-2.

This proposal might be useful to answer research questions like these:

> Can MEXICA-impro represent the refugees and residents' clashing perspectives?
>
> Will MEXICA-impro generate narratives that include the concerns and hopes of both, refugees and residents, in a coherent piece of text?
>
> Can this type of narrative contribute to a better understanding between people?
>
> Can a machine capable of generating narratives where diverse perspectives converge function as a motivating element for refugees and residents to listen to each other? Or are these types of products irrelevant for the targeted audience?

And so on.

14.2.4 What about education?

There is currently an intense debate about the effects that AI will have on education. One of the main concerns between teachers is that students will use systems like transformers to generate their written assignments instead of writing those texts themselves.[3] If an AI program can answer an essay question, perhaps the time has come to reflect on whether this is the best way to assess our students. On the other hand, narrative-generating systems can be useful tools for promoting critical thinking. For example, imagine that a teacher generates essays with the help of AI, and then asks his students to critique these texts and write an improved version. Or that this teacher asks his students to design a test with several questions on some topic, then the students use an AI system to generate the answers, and, finally, they grade them (Sharples 2022). These and other similar exercises can be of great help in the training of our students.

These examples illustrate the potential of automatic narrative generators to contribute to the development of projects useful for society. A critical AI society must be attentive to unforeseen situations that arise as a result of the advances of these technologies, and thus be able to estimate their impact on society.

14.3 The end

Automatic narrative generation is a fascinating topic that comprises scientific, technological, social, economic, and cultural dimensions. This book studies core features of the scientific and technological dimensions, and reflects on some of their consequences for society. There are still important questions related to the other dimensions that need to be answered. For instance:

> Can automatic narrative generators be massively commercialized? Will these technologies substitute for human jobs? Will these technologies increase the gap between rich and poor?
> Can these technologies facilitate the development of new skills? Can they contribute to the development of literature?
> Can automatic narrative generation give a voice to under-represented communities?

And so on.

A research project that focuses on the interaction of all the dimensions mentioned will provide a comprehensive panorama suitable for answering these and other related questions.

---

[3] Mike Sharples and Rafael Pérez y Pérez, Original essays written in seconds: how 'transformers' will change assessment. *Times Higher Education* (4 July 2022). Available at https://www.timeshighereducation.com/campus/original-essays-written-seconds-how-transformers-will-change-assessment

We have reached the end of this book. It narrates the story we have always wanted to tell—a story where we argue about the importance of reducing the AI knowledge gap to develop a critical AI society, a story that explores the intimacy of narrative generators. Our aim throughout the text has been to show how computer models could profoundly change the ways we see the process of writing and constructing ourselves through stories. We hope we have achieved this goal. We are certain that the future will bring lots of surprises. Perhaps the next edition of this book will include as a co-author an automatic narrative generator. We are looking forward to it.

References

Aggarwal, C. C. (2018). *Neural networks and deep learning: a textbook*. Cham: Springer.

Alhussain, A. I., and Azmi, A. M. (2021). Automatic story generation: a survey of approaches. *ACM Computing Surveys*, 54(5), 1–38. https://doi.org/10.1145/3453156

Appelbaum. M. A. (1976). *Meta-symbolic simulation system (MESSY): user manual*. Technical Report 272. Computer Science Department, University of Wisconsin.

Bae, B. C., and Young, M. (2014). A computational model of narrative generation for surprise arousal. *IEEE Transactions on Computational Intelligence and AI in Games*, 6(2), 131–143.

Bahdanau, D., Cho, K., and Bengio, Y. (2014). Neural machine translation by jointly learning to align and translate. https://arxiv.org/abs/1409.0473

Basharina, G. P., Langvilleb, A. N., and Naumov, V. A. (2004). The life and work of A.A. Markov. *Linear Algebra and its Applications*, 386, 3–26.

Beale, R., and Jackson, T. (1992). *Neural computing: an introduction*. Bristol: Institute of Physics Publishing.

Becker, A. (2006). A review of writing model research based on cognitive processes. In: Horning, A., and Becker, A. (eds), *Revision: history, theory, and practice*, pp. 25–49. West Lafayette, IN: Parlor Press.

Bender, E. M., Gebru, T., McMillan-Major, A., and Shmitchell, S. (2021). On the dangers of stochastic parrots: can language models be too big? In: Proceedings of the 2021 ACM Conference on Fairness, Accountability, and Transparency, pp. 610–23, https://doi.org/10.1145/3442188.3445922

Black, J. B., and Wilensky, R. (1979). An evaluation of story grammars. *Cognitive Science*, 3, 213–30.

Brown, T. B., Mann, B., Ryder, N., Subbiah, M., Kaplan, J., Dhariwal, P., et al. (2020). Language models are few-shot learners. arXiv: 2005.14165v1

Clayton, J. J. (1996). Introduction: on fiction. In *The Heath introduction to fiction*, pp. 1–32. Lexington, MA: D.C. Heath and Company.

Compton, K., Kybartas, B., and Mateas, M. (2015). Tracery: an author-focused generative text tool. In: Schoenau-Fog, H., Bruni, L., Louchart, S., Baceviciute, S. (eds), *Interactive storytelling*. ICIDS 2015. Lecture Notes in Computer Science, vol 9445, pp. 154–61. Cham, Switzerland: Springer. https://doi.org/10.1007/978-3-319-27036-4_14

D'Avila Garcez, A. S., and Lamb, L. C. (2020). Neurosymbolic AI: the 3rd wave. arXiv: 2012.05876

Deane, P., Odendahl, N., Quinlan, T., Fowles, M., Welsh, C., and Bivens-Tatum, J. (2008). Cognitive models of writing: writing proficiency as a complex integrated skill. *ETS Research Report Series*, 2008, i–36.

Dostoyevsky, F. (2009). *The Brothers Karamazov*. Urbana, IL: Project Gutenberg. Retrieved 17 March 2020, from http:www.gutenberg.org/ebooks/28054. Translated by Garnett Constance in 1912.

Evans, J. S. B. (2003). In two minds: dual-process accounts of reasoning. *Trends in Cognitive Sciences*, 7(10), 454–9.

Exploring the fascinating world of language. *Business Machines* 46(3), 10–11 (1963), courtesy IBM Corporate Archives.

Fan, A., Lewis, M., and Dauphin, Y. (2018). Hierarchical neural story generation. In: Proceedings of the 56th Annual Meeting of the Association for Computational Linguistics, pp. 889–98. Association for Computational Linguistics.

Fauconnier, G., and Turner, M. (2002). *The way we think: conceptual blending and the mind's hidden complexities*. New York: Basic Books.

Ferretti, F., Adornetti, I., Chiera, A., Nicchiarelli, S., Magni, R., Valeri, G., et al. (2017). Mental time travel and language evolution: a narrative account of the origins of human communication. *Language Sciences*, 63, 105–18.

Gelernter, D. (1994). *The muse in the machine*. London: Fourth Estate.

Genette, G. (1980). *Narrative discourse: an essay in method*, trans. Jane E. Lewin. Ithaca, NY: Cornell University Press.

Genette, G. (1988). *Narrative discourse revisited*, trans. Jane E. Lewin. Ithaca, NY: Cornell University Press.

Gervás, P. (2009). Computational approaches to storytelling and creativity. *AI Magazine*, 49–62.

Gervás, P. (2012). From the fleece of fact to narrative yarns: a computational model of narrative composition. In: *Proceedings of* The Third Workshop on Computational Models of Narrative *(CMN'12)*, pp. 125–33. http://narrative.csail.mit.edu/cmn12/proceedings.pdf

Gervás, P. (2014). Composing narrative discourse for stories of many characters: A case study over a chess game. *Literary and Linguistic Computing*, 29(4), 511–31. https://doi.org/10.1093/llc/fqu040

Goguen, J., and Harrell, D. F. (2009). Style as a choice of blending principles. In: Argamon, S., Burns, K., and Dubnov, S. (eds), *The structure of style: algorithmic approaches to understanding manner and meaning*, pp. 291–317. Berlin: Springer.

Goodfellow, I., Bengio, Y., and Courville, A. (2106). *Deep learning (adaptive computation and machine learning series)*. Cambridge, MA: MIT Press.

Grimes, J. E. (1965). La computadora en las investigaciones humanísticas. *Anuario de Letras. Lingüística y Filología*, 5, 163–74.

Harrell, D. F. (2005). Shades of computational evocation and meaning: the GRIOT system and improvisational poetry generation. In: *Proceedings of* the 6th Digital Arts and Culture Conference *(DAC 2005), Copenhagen, Denmark*, pp. 133–43.

Harrell, D. F. (2007). *Theory and technology for computational narrative: an approach to generative and interactive narrative with bases in algebraic semiotics and cognitive linguistics*. PhD dissertation, Department of Computer Science and Engineering, University of California, San Diego, La Jolla.

Harrell, F. (2013). *Phantasmal media. an approach to imagination, computation, and expression*. Cambridge, MA: MIT Press.

Herrera-González, B. D., Gelbukh, A., and Calvo, H. (2020). Automatic story generation: state of the art and recent trends. In: Martínez-Villaseñor, L., Herrera-Alcántara, O., Ponce, H., and Castro-Espinoza, F. A. (eds), *Advances in Computational Intelligence: MICAI 2020. Lecture Notes in Computer Science*, vol. 12469, pp. 81–91. Cham: Springer. https://doi.org/10.1007/978-3-030-60887-3_8

Hochreiter, S., and Schmidhuber, J. (1997). Long short-term memory. *Neural Computation* 9(8), 1735–80. DOI: 10.1162/neco.1997.9.8.1735

Hou, C., Zhou, C., Zhou, K., Sun, J., and Xuanyuan, S. (2019). A survey of deep learning applied to story generation. In: Qiu, M. (ed.) *Smart Computing and Communication. SmartCom 2019. Lecture Notes in Computer Science*, vol 11910, pp. 1–10. Cham: Springer. https://doi.org/10.1007/978-3-030-34139-8_1

Jordan, M. I., and Russell, S. (1999). Computational intelligence. In: Wilson, R. A., and Keil, F. C. (eds), *The MIT encyclopedia of the cognitive sciences*, pp. lxxiii–xc. Cambridge, MA: MIT Press.

Klein, S., Aeschlimann, J. F., Balsiger, D. F., Converse, S.L., Court, C., Foster, M., et al. (1973). *Automatic novel writing: a status report*. Technical Report 186. Computer Science Department, University of Wisconsin.

Kotseruba, I., and Tsotsos, J. K. (2020). 40 years of cognitive architectures: core cognitive abilities and practical applications. *Artificial Intelligence Review* 53, 17–94.

Kybartas, B., and Bidarra, R. (2017). A survey on story generation techniques for authoring computational narratives. *IEEE Transactions on Computational Intelligence and AI in Games*, 9(3), 239–53.

Lakoff, G. P. (1972). Structural complexity in fairy tales. *The Study of Man*, 1, 128–90.

Läufer, M. (2018). *A noise such as a man might make.* Denver, CO: Counterpath Press.

Läufer, M. (2018). *Scenes from a marriage.* http://www.miltonlaufer.com.ar/256nano/scenesfromamarriage.php, accessed 6 December 2019.

Lebowitz, M. (1985) Story-telling as planning and learning. *Poetics* 14, 483–502.

León, C., Gervás, P., and Delatorre, P. (2019). Empirical insights into short story draft construction. IEEE Access, 7, 119192–208. DOI: 10.1109/ACCESS.2019.2936919.

Lodge, D. (1996). *The practice of writing: essays, lectures, reviews and a diary.* London: Secker & Warbug.

McCulloch, W., and Pitts, W. (1943). A logical calculus of the ideas immanent in nervous activity. *Bulletin of Mathematical Biophysics*, 5, 115–133 (reprinted, 1990, 52(1/2), 99–115).

Mar, R. A., Li, J., Nguyen, A. T., and Ta, C. P. (2021). Memory and comprehension of narrative versus expository texts: a meta-analysis. *Psychonomic Bulletin & Review*, 1–18.

Marcus., G. (2020). The next decade in AI: four steps towards robust artificial intelligence. https://doi.org/10.48550/arXiv.2002.06177

Martin, L. J. (2021) *Neurosymbolic automated story generation.* PhD dissertation, School of Interactive Computing, Georgia Institute of Technology.

Martin, L. J., Ammanabrolu, P., Wang, X., Hancock, W., Singh, S., Harrison, B., et al. (2017). Event representations for automated story generation with deep neural nets. arXiv: 1706.01331v3.

Mateas, M. (2002). *Interactive drama, art and artificial intelligence.* PhD dissertation, Carnegie Mellon University.

Meehan, J. R. (1976). *The metanovel: writing stories by computer.* PhD dissertation, Yale University.

Mikolov, T., Chen, K., Corrado, g., and Dean, J. (2013). Efficient estimation of word representations in vector space. arXiv: 1301.3781v3

Minsky, M., & Papert, S. (1969). *Perceptrons.* Cambridge, MA: MIT Press.

Montfort, N. (2007a). *Generating narrative variation in interactive fiction.* PhD dissertation, University of Pennsylvania. http://nickm.com/if/Generating_Narrative_Variation_in_Interactive_Fiction.pdf

Montfort, N. (2007b). The simulated bank robbery. Available at https://nickm.com/curveship/js/examples/robbery.html, accessed 7 November 2020.

Montfort, N. (2011). Curveship's automatic narrative style. In: *FDG '11: Proceedings of the 6th International Conference on Foundations of Digital Games, June 2011*, pp. 211–18.

Montfort, N., and Pérez y Pérez, R. (2008). Integrating a plot generator and an automatic narrator to create and tell stories. In: *Proceedings of the 5th International Joint Workshop on Computational Creativity, Universidad Complutense de Madrid, España*, pp. 61–70.

Montfort, N., Pérez y Pérez R., Harrell, F., and Campana, A. (2013). Slant: a blackboard system to generate plot, figuration, and narrative discourse aspects of stories. In: Proceedings of the Fourth International Conference on Computational Creativity, *Sydney, Australia*, pp. 168–75.

Mueller, E. T. (1987). *Daydreaming and computation: a computer model of everyday creativity, learning, and emotions in the human stream of thought* (Technical Report UCLA-AI-87-8). Doctoral dissertation, Computer Science Department, University of California, Los Angeles.

Mueller, E. T. (1990). *Daydreaming in humans and machines: a computer model of the stream of thought*. Norwood, NJ: Ablex Publishing.

Nye, M., Tessler, M. H., Tenenbaum, J. B., and Lake, B. M. (2021). Improving coherence and consistency in neural sequence models with dual-system, neuro-symbolic reasoning. arXiv:2107.02794v2

Nystrand, M. (2006). The social and historical context for writing research. In: MacArthur, C. A., Graham, S., and Fitzgerald, J. (eds), *Handbook of writing research*, pp. 11–27. New York: Guilford Press.

Ontañón, S., and Zhu, J. (2010). Story and text generation through computational analogy in the Riu system. In: Proceedings of the Sixth AAAI Conference on Artificial Intelligence and Interactive Digital Entertainment, pp. 51–6.

Pemberton, L. (1989). A modular approach to story generation. In: Proceedings of Fourth Conference of the European Chapter of the Association for Computational Linguistics, *Manchester, UK*, pp. 217–24.

Pérez y Pérez, R. (1999). *MEXICA: a computer model of creativity in writing*. DPhil dissertation, University of Sussex, UK.

Pérez y Pérez, R. (2007). Employing emotions to drive plot generation in a computer-based storyteller. *Cognitive Systems Research*, 8(2), 89–109. DOI: 10.1016/j.cogsys.2006.10.001

Pérez y Pérez, R. (2014). The three layers evaluation model for computer-generated plots. In: *Proceedings of the Fifth International Conference on Computational Creativity, Ljubljana, Slovenia*, pp. 220–9. http://computationalcreativity.net/iccc2014/proceedings/

Pérez y Pérez, R. (2015a). Reflexiones sobre las características del trabajo interdisciplinario y sugerencias sobre cómo fomentarlo en el aula universitaria. In: Castellanos, V. (ed.) *Estudios Interdisciplinarios en Comunicación*, pp. 33–50, México: UAM Cuajimalpa. http://www.casadelibrosabiertos.uam.mx/contenido/contenido/Libroelectronico/Estudios-interdisciplinarios.pdf

Pérez y Pérez, R. (2015b). A computer-based model for collaborative narrative generation. *Cognitive Systems Research*, 36–7, 30–48. http://dx.doi.org/10.1016/j.cogsys.2015.06.002

Pérez y Pérez, R. (2015c). MEXICA-impro: Generación automática de narrativas colectivas. In: Pérez y Pérez, R. (ed.), *Creatividad Computacional*, UAM-Cuajimalpa-Patria, pp. 95–110.

Pérez y Pérez, R. (2017). *MEXICA: 20 years–20 stories [20 años–20 historias]*. Denver, CO: Counterpath Press.

Pérez y Pérez, R. (2019). Representing social common sense knowledge in MEXICA. In: Veale, T., and Cardoso, A. (eds), *Computational creativity: the philosophy and engineering of autonomously creative systems*, pp. 255–74. Cham, Switzerland: Springer.

Pérez y Pérez, R., and Sharples, M. (2001). MEXICA: a computer model of a cognitive account of creative writing. *Journal of Experimental and Theoretical Artificial Intelligence*, 13(2), 119–39. DOI: 10.1080/09528130118867

Pérez y Pérez, R., and Sharples, M. (2004). Three computer-based models of storytelling: BRUTUS, MINSTREL and MEXICA. *Knowledge Based Systems Journal*, 17(1), 15–29.

Prince, G. (1980). Aspects of a grammar of narrative. *Poetics Today*, 1(3), 49–63.

Prince, G. (1982). *Narratology: the form and functioning of narrative*. Berlin: Mouton.

Prince, G. (1988). The disnarrated. *Style*, 22(1), 1–8.

Propp, V. (2010). *Morphology of the folktale*. Austin: University of Texas Press (originally published in 1928).

Quoidbach, J., and Dunn, E. W. (2013). Affective forecasting. In: Pashler, H. (ed.), *Encyclopedia of the mind*, pp. 11–13. Los Angeles: SAGE.

Rich, E., and Knight, K. (1991). *Artificial intelligence*, 2nd edn. New York: McGraw-Hill.

Riedl, M. O. (2004). *Narrative generation: balancing plot and character*. PhD thesis, North Carolina State University.

Rosenblatt, F. (1962). *Principles of neurodynamics: perceptrons and the theory of brain mechanisms*. Washington, DC: Spartan Books.

Rumelhart, D. E., and McClelland, J. L. (1986). *Parallel distributed processing*, Vol. 1. Cambridge, MA: MIT Bradford Press.

Ryan, J. (2017). Grimes' fairy tales: a 1960s story generator. In: Proc. International Conference on Interactive Digital Storytelling, pp. 89–103. Cham: Springer.

Ryan, J. (2018). Curating simulated storyworlds. PhD dissertation, University of California Santa Cruz. Available at https://escholarship.org/content/qt1340j5h2/qt1340j5h2.pdf

Santoro, A., Lampinen, A., Mathewson, K., Lillicrap, T., and Raposo, D. (2022). Symbolic behaviour in artificial intelligence. arXiv: 2102.03406v2

Seidl, J., and McMordies, W. (1991). *English idioms*, 5th edn. Oxford, UK: Oxford University Press.

Sharples, M. (1999). *How we write? Writing as creative design*. London: Routledge.

Sharples, M. (2022). Automated essay writing: an AIED opinion. *International Journal of Artificial Intelligence in Education*, 32, 1119–26.

Sharples, M., and Pérez y Pérez, R. (2022). *Story machines. how computers have become creative writers*. London: Routledge.

Smith, D., Schlaepfer, P., Major, K., Dyble, M., Page, A. E., Thompson, J., et al. (2017). Cooperation and the evolution of hunter-gatherer storytelling. *Nature Communications*, 8(1), 1–9.

Strachey, C. (1954). The 'thinking' machine. *Encounter*, 25–31.

Tambwekary, P., Dhuliawalay, M., Martin, L. J., Mehtay, A., Harrisonz, B., and Riedly, M. O. (2019). Controllable neural story plot generation via reward shaping. arXiv: 1809.10736v3

Thorndyke, P. W. (1977). Cognitive structures in comprehension and memory of narrative discourse. *Cognitive Psychology*, 9, 77–110.

Turner, M. (2015). La integración conceptual como un programa de investigación para la creatividad computacional. In: Pérez y Pérez, R. (ed.), *Creatividad Computacional*, pp. 33–50. Ciudad de México: Grupo Editorial Patria.

Turner, S. R. (1993). *MINSTREL: A computer model of creativity and storytelling*. PhD dissertation, University of California, Los Angeles.

Turner, S. R. (1994). *The creative process: a computer model of storytelling*. Hillsdale, NJ: Lawrence Erlbaum Associates.

Vaswani, A., Shazeer, N., Parmar, N., Uszkoreit, J., Jones, L., Gomez, A. N., Kaiser, L., et al. (2017). Attention is all you need. arXiv: 1706.03762v5.

Veale, T. (2016). Round up the usual suspects: knowledge-based metaphor generation. In: Proceedings of the Fourth Workshop on Metaphor in NLP, *San Diego, California*, pp. 34–41. Association for Computational Linguistics. DOI: 10.18653/v1/W16-1105

Veale, T. (2017). Déjà vu all over again. on the creative value of familiar elements in the telling of original tales. In: Proceedings of the Eight International Conference on Computational Creativity, Atlanta, pp. 245–52.

Veale, T. (2021). *Your wit is my command: building AIs with a sense of humor*. Cambridge, MA: MIT Press.

Wallas, G. (1926) *The art of thought*. London: Butler & Tanner. https://archive.org/details/theartofthought

Wei, J., Tay, Y., Bommasani, R., Raffel, C., Zoph, B., Borgeaud, S., et al. (2022). Emergent abilities of large language models. arXiv: 2206.07682.

Weisberg, R. W. (1986). *Creativity: genius and other myths*. New York: W. H. Freeman.

Weizenbaum, J. (1966). ELIZA—a computer program for the study of natural language communication between man and machine. *Communications of the ACM* 9(1), 36–45.

Wilson, R. A., and Keil, F. C. (1999) (eds). *The MIT encyclopedia of the cognitive sciences*. Cambridge, MA: MIT Press.

Zhang, S., Roller, S., Goyal, N., Artetxe, M., Chen, M., Chen, S., et al. (2022). OPT: open pretrained transformer language models. arXiv: 2205.01068.

Index

For the benefit of digital users, indexed terms that span two pages (e.g., 52–53) may, on occasion, appear on only one of those pages.

Tables are indicated by an italic *t* following the page/paragraph number.